"From John Calvin to Wendell Berry and Elon M[...] to medical innovations and space travel. From t[...] lem. From technology as savior to the sovereign[...] Valley to the blood of Jesus. Tony Reinke's book *God, Technology, and the Christian Life* is panoramic and penetrating. I doubt there has ever been a more sweeping treatment of technology so firmly tethered to Scripture—and therefore so realistic and hopeful. Writing as a 'tech optimist' who trusts in God's providential orchestration over all things, Reinke offers us an expansive and compelling 'biblical theology of technology.' God's glory is the end of creation and the aim of all innovations. Apart from Christ there is no art, no science, no technology, no agriculture, no microprocessor, and no medical innovation. If God is the center of our life, technology is a great gift. If technology is our savior, we are lost. This is a mind-expanding, heart-stabilizing, God-glorifying, joy-sustaining book."

John Piper, Founder and Teacher, desiringGod.org; Chancellor, Bethlehem College & Seminary; author, *Desiring God*

"Tony Reinke has written a must-read for any Christian seeking to understand God's view of technology. *God, Technology, and the Christian Life* masterfully fuses together scriptural commentary, historical wisdom, and practical application to give a rich, Christian worldview of technology. A more positive view of human innovation and innovators is a breath of fresh air in a cultural moment when technology is viewed more as a harm than a help to many Christians. I will be recommending this to our FaithTech community worldwide."

James Kelly, Founder and CEO, FaithTech

"Reinke not only addresses a wide range of issues in technology and culture; he also brings fresh insights into often overlooked passages of Scripture. He offers an approach to technology that is ethical without being moralistic, careful without being restrictive, and positive without being naive."

John Dyer, Dean and Professor, Dallas Theological Seminary; author, *From the Garden to the City: The Place of Technology in the Story of God*

"As both a pastor and an engineer, I continually find the need to interpret the marvels of the twenty-first century in light of Scripture. To that end, this book has been a great blessing. Tony Reinke has crafted an enlightening, balanced, and thoroughly engaging biblical theology of technology. This work is profoundly practical. All Christians should consider it, whether they live inside a major tech center or not."

Conley Owens, Pastor, Silicon Valley Reformed Baptist Church; Senior Engineer, Google

"Tony has given us a rich suite of resources for the believer who wishes to make sense of technology's increased role in society and in our individual lives. This is not a fear-based, hasty string of reactionary warnings, but a careful look at the complex, intimate, and unavoidable relationship between technology and theology. We are ably guided through a detailed, God-centered tour of the history of technology, from Babel to Bumble, using theologians, inventors, and philosophers. Take advantage of this excellent work."

Jared Oliphint, Philosophy Department, Texas A&M University

"*God, Technology, and the Christian Life* is a dangerous read for the serious-minded believer. Here Reinke unearths the source of all technology from the very pages of Scripture, forcing the Christian to view this evolving fixture of the modern world through the curative lens of a sovereign God and the unfading hope of the gospel. Whether cynical or exhilarated by the breakneck speed of innovation in the twenty-first century, this book will challenge how we all see and interact with our ever-changing world."

Jeremy Patenaude, Pastor, Risen Hope Church, Seattle; writer, Microsoft

"The story of God's glory is still unfolding inside the zeitgeist of the technium. Whether talking about developers inventing new apps in a data center, automated manufacturing robots churning out electric cars, or the eager consumers of these new products and services, this book reminds us that human technology serves God's final purpose for his creation. In this captivating book, Tony offers an optimistic theology of technology that will inspire us to worship the Creator of our most powerful inventors, and—astonishingly so—help us live cautiously and faithfully inside our technological cities. To do it, he demystifies concepts created by well-intentioned Christians over the decades who have made it hard to see that science and technology exist by God, through God, and for God. His glory is reflected in ammonia, lithium, nuclear fission, and in advances to come in nuclear fusion and space travel. *God, Technology, and the Christian Life* is essential reading for pastors, church leaders, and every Christian who lives and works inside the technological cities of man. A wake-up call for us to anticipate Christ's return and the arrival of a new city—a better city—designed and built by God himself."

Jose Luis Cuevas, pastor; missionary; Director of Project Management, Office for VMware Inc., Latin America

"Given the acceleration of automation in every aspect of our lives, we all need to reflect deeply on our technology history and future roadmap. In *God, Technology, and the Christian Life*, Tony Reinke has developed a gospel-centered analysis of our technology-driven culture that is beneficial for both Christians and non-Christians alike."

Bernie Mills, Vice President, VMware Inc.; Board Member, Joni and Friends

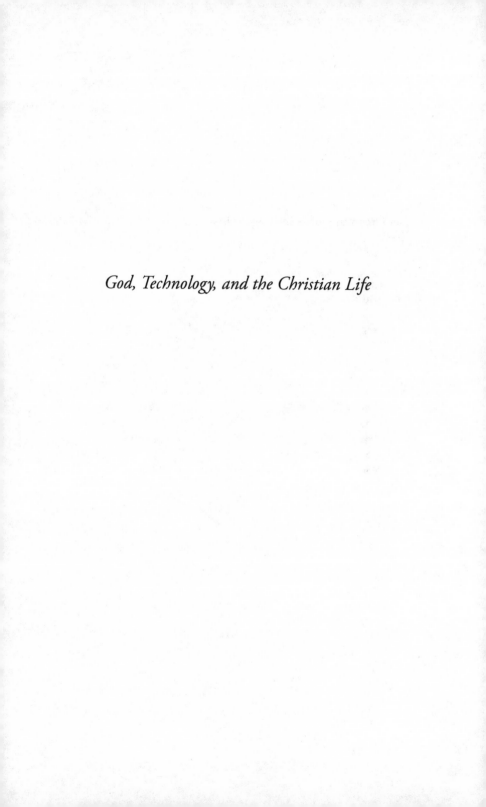

God, Technology, and the Christian Life

Other Crossway Books by Tony Reinke

Competing Spectacles: Treasuring Christ in the Media Age (2019)

Lit! A Christian Guide to Reading Books (2011)

Newton on the Christian Life: To Live Is Christ (2015)

12 Ways Your Phone Is Changing You (2017)

God, Technology, and the Christian Life

Tony Reinke

WHEATON, ILLINOIS

Library of Congress Cataloging-in-Publication Data
Names: Reinke, Tony, 1977- author.
Title: God, technology, and the Christian life / Tony Reinke.
Description: Wheaton, Illinois : Crossway, 2022. | Includes bibliographical references and index.
Identifiers: LCCN 2021006950 (print) | LCCN 2021006951 (ebook) | ISBN 9781433578274 (trade paperback) | ISBN 9781433578281 (pdf) | ISBN 9781433578298 (mobipocket) | ISBN 9781433578304 (epub)
Subjects: LCSH: Technology—Religious aspects—Christianity.
Classification: LCC BR115.T42 R46 2022 (print) | LCC BR115.T42 (ebook) | DDC 261.5/6—dc23
LC record available at https://lccn.loc.gov/2021006950
LC ebook record available at https://lccn.loc.gov/2021006951

Crossway is a publishing ministry of Good News Publishers.

LSC			31	30	29	28	27	26	25	24	23	22
15	14	13	12	11	10	9	8	7	6	5	4	3	2	1

*Dedicated to every Christian living inside demanding and
expensive tech centers, unselfishly building churches, and
influencing the world's most powerful industries for good.*

We have detailed studies about the history of individual technologies
and how they came into being. We have analyses of the design
process; excellent work on how economic factors influence the
design of technologies, how the adoption process works, and how
technologies diffuse in the economy. We have analyses of how
society shapes technology, and of how technology shapes society.
And we have meditations on the meaning of technology, and
on technology as determining—or not determining—human
history. But we have no agreement on what the word "technology"
means, no overall theory of how technologies come into being,
no deep understanding of what "innovation" consists of, and
no theory of evolution for technology. Missing is a set of overall
principles that would give the subject a logical structure, the sort
of structure that would help fill these gaps. Missing, in other
words, is a theory of technology—an "-ology" of technology.

W. BRIAN ARTHUR

Contents

1

What Is Technology?

PEOPLE DON'T SLOW DOWN much when driving across Nebraska. But tap the brakes the next time you're crossing the Cornhusker state, glance off into a cornfield, and you might see my name in all caps. REINKE is synonymous with agricultural technology. The name swings on metal logos attached to giant farm sprinklers across the Midwest because my grandfather and his five brothers claimed three dozen patents among them for ideas ranging from the aspirational to the multimillion-dollar useful.[1] The ideas that paid off seeded a corporation of center-pivot irrigation systems for farms and aluminum truck beds for semis.

My grandfather's technological ambition was undampened by a lack of schooling past eighth grade. A carpenter, electrician, and farmer, he was awarded a bronze star in World War II for helping to reengineer an antiaircraft aiming computer.[2] Back home, he

---

1 Susan Harms, "Area Reinkes Are Brothers of Invention," *Hastings Daily Tribune* (n.d.), n.p.
2 I know little detail of the "M5A1 Director" 40mm gun directing/aiming device beyond the description in Captain Kirby M. Quinn, "Gunning for War Birds," *Popular Mechanics*, December 1933, 801–4.

aspired to modernize rural homesteads, turning hundred-year-old houses built prior to running water into electrified homes powered by batteries recharged by aluminum windmills. In his personal machine shop, he invented and manufactured copper heat exchangers to cool irrigation engines with groundwater.

When electrical costs soared in 1978, my grandfather designed and built an aluminum windmill using a centripetal flywheel to automatically pitch the blades based on wind speed, making it possible to generate electricity with either high wind or very little.[3] He was fascinated by aluminum. For fun, he crafted the first aluminum violin I have ever seen (and thankfully the last one I have ever heard).[4] By the time my grandfather retired, I was in high school, and he cleaned out his workshop by giving me a pile of abandoned aluminum projects. It took me weeks to pneumatically chisel thousands of aluminum rivets off iron structures, but it paid off. By the end of that summer, the pile of broken rivets and the sheets of scrap metal registered into an aluminum pile that I recycled for one thousand dollars. It helped pay for college. But more memorably, it put me in close proximity to the remnants of my grandfather's ambitious dreams.

Innovation is in the Reinke blood. But technology is deeply connected to each of us. The story of humanity is the story of technology. The prophet Daniel marked off successive kingdoms by dominant metals: gold, silver, bronze, iron, iron-clay.[5] We mark

---

3 Claire Hurlbert, "Davenport Man Plugs into Nebraska's Wind Power," *Hastings Daily Tribune*, August 26, 1978, n.p.

4 He apparently was not alone. Aluminum bass violins were featured in *Popular Mechanics* (December 1933) boasting of "a tonal quality comparing favorably with the finest wood basses" (805). I harbor doubts, but aluminum instruments were a cultural aspiration in his era.

5 Dan. 2:31–45.

off human history by the Stone Age, the Bronze Age, the Iron Age, the nuclear age, and the computer age. Today, we live in the age of technology. This long-running drama of innovation includes each of us. No family tree is uninventive.

This Reinke lives inside an accelerating tech age that the world has never seen. I don't think my grandfather ever touched a PC, but someday I may be biologically linked to a superprocessor. My father, himself very inventive, was mesmerized by the moon landing. But in my lifetime, I expect to see commercial flights to the moon. Right now, I could spit saliva in a tube, mail it, and get a full mapping of my heredity and genetic susceptibilities. My great grandkids may live on Mars. I have witnessed incredible changes in my first forty years on this planet and, Lord willing, I brace for more outrageous changes to come in the forty years ahead—or century ahead, if the prophets of life expectancy are right.

I don't innovate in a farm shop like my grandfather; I write in the outskirts of a major city, surrounded by technology. As I type, my robot vacuum bumps my feet, stops, turns, bumps, stops, turns, and bumps again, self-correcting like a blind turtle as it cleans the carpet in my office. Specialized automated robots, like my vacuum droid, can do one thing well, but nothing else. Remote-controlled bomb-detonating robots sync with other semiautonomous service robots. Prototype dog-like robots and human-like robots are in the works at major science labs. And at the far ends of the robotics industry are inhuman sex robots and weaponized killing robots. And the first-ever, fully autonomous robots are likely to appear in the next several years. We call them driverless cars.

We are entering a new technological revolution that's impossible to predict. It's a good time for Christians to think about God's relationship to technology as we ask questions about the origin of our gadgets. What technologies are helpful or destructive? And how can we walk by faith in the age ahead? First, we must tackle a fundamental question: What is technology?

What Is Technology?

Technology is applied science and amplified power. It's art, method, know-how, formulas, and expertise. The word *technology* is built on the root *techne-* or *technique*. We amplify our native powers through new techniques. Noah and the animals could never outswim a global flood, so God designed a ship. The people of Babel couldn't live in the sky, so they engineered a tower. Today, elevators in downtown Dubai carry people into the stratosphere. Jacob and his sons dug wells by hand and shovel, but Union Pacific blasted trails through the mountains with dynamite. Today, dinosaur-sized augers grind out underground tunnels for millions of telecom cables. And the smartphone extends the popping electrical explosions in our brains, through our thumbs, to our phones to become little digital ones and zeroes that we broadcast in messages to influence the world.

Tech intensifies our dexterity, augments our influence, and empowers our previously feeble intentions. And no innovation more potently amplifies us like the computer chip. By weight these little chips are the most powerful things in the continuous universe. Excluding cosmic explosions and nuclear bombs that exhaust their power in a hyperblink, "of all the sustainable things in the universe, from a planet to a star, from a daisy to an automobile, from a brain to an eye, the thing that is able to conduct the highest density of power—the most energy flowing through a gram of matter each second—lies at the core of your laptop." Yes, the tiny microprocessor "conducts more energy per second per gram through its tiny corridors than animals, volcanoes, or the sun." The computer chip is "the most energetically active thing in the known universe."[6]

As I write, Apple has just unveiled M1, "the most powerful chip" the company has ever made, "packed with an astounding 16 billion transistors."[7] With this much power in every iPhone and MacBook, we can do a lot with our tools—a lot of damage or a lot of good. So how will we wield this power?

Learned techniques are ancient too. When the Good Samaritan found a bleeding Jew on the street, he jumped into action, binding the wounds and applying topical treatments before loading the man's weight like cargo on his animal and transporting him to an inn where he paid with money he made in the market so the innkeeper would continue the work of applying healing measures.[8] The story shows us love in action through technique. We don't love "by smiling in abstract beneficence on our neighbors," wrote agrarian Wendell Berry. No, our love "must come to acts, which must come from skills. Real charity calls for the study of agriculture soil husbandry, engineering, architecture, mining, manufacturing, transportation, the making of monuments and pictures, songs and stories. It calls not just for skills but for the study and criticism of skills, because in all of them a choice must be made: they can be used either charitably or uncharitably."[9] We love one another through art, skill, and technology.

The story of humanity tells the tale of how we have learned to love each other more by improving our skills. Back in the fifth century, Augustine pondered all the ways that we use our talents

6   Kevin Kelly, *What Technology Wants* (New York: Penguin, 2011), 59–60.
7   "Apple Unleashes M1," press release, apple.com (Nov. 10, 2020).
8   Luke 10:30–37.
9   Wendell Berry, *Essays 1969–1990*, ed. Jack Shoemaker (New York: Library of America, 2019), 525.

to serve society. He praised the intellect of fallen sinners, the intact "natural genius of man," that creates remarkable necessary inventions (and unnecessary ones too). When making a list of innovations that caught his attention, Augustine began with textiles, architecture, agriculture, and navigation. Then he celebrated sculptors, painters, composers, and theater producers. Then he turned his attention to nature, and all the ways humans capture, kill, or train wild animals. Then he thought of all the medical drugs that preserve and restore human health, without forgetting the weapons used to defend one's country in war. Next, he praised the "endless variety of condiments and sauces which culinary art has discovered to minister to the pleasures of the palate." (Translation: give thanks for Chick-fil-A Sauce.) Next, he commended all the means we have created for speaking and writing and communicating, from rhetoric and poems to novels and lyrics. And then he praised musicians with instruments and songs. Mathematicians next. Then astronomers. For Augustine, you can pick any branch of science, follow its course, and be captivated by human ingenuity. Over every imaginative invention of man we celebrate "the Creator of this noble human nature" who is "the true and supreme God whose providence rules all that he has created."[10]

Everything mentioned here by Augustine (down to sauces), includes applied science, or *technology*. In 1829 Jacob Bigelow published a book with that relatively new term in the title: *Elements of Technology*, a book to celebrate advances in human writing, painting, sculpting, architecture, building, heating, ventilation, lighting, wheels, machines, textiles, metallurgy, and food preservation. All

---

10 Augustine of Hippo, *The City of God*, bks. 17–22, ed. Hermigild Dressler, trans. Gerald G. Walsh and Daniel J. Honan, vol. 24, *The Fathers of the Church* (Washington, DC: Catholic University of America Press, 1954), 484–85.

of these advances are *technology*—"a word sufficiently expressive, which," he said, "is beginning to be revived in the literature of practical men at the present day."[11]

It caught on. *Technology* is now a household term for all the tools we wield. We innovate through skills. We make new techniques. Technology is essential to who we are, in every era—from the age of the semiautomatic rifle to the age of the slingshot.

A Famous Tech Story

Our technologies can be primitive or advanced, a distinction that reminds me of the story of David and Goliath, two technologists who clashed in 1 Samuel 17:4–40. Here's how the story begins, with advanced weaponry in verses 4–11.

> [4]And there came out from the camp of the Philistines a champion named Goliath of Gath, whose height was six cubits and a span. [5] He had a helmet of bronze on his head, and he was armed with a coat of mail, and the weight of the coat was five thousand shekels of bronze. [6] And he had bronze armor on his legs, and a javelin of bronze slung between his shoulders. [7] The shaft of his spear was like a weaver's beam, and his spear's head weighed six hundred shekels of iron. And his shield-bearer went before him. [8] He stood and shouted to the ranks of Israel, "Why have you come out to draw up for battle? Am I not a Philistine, and are you not servants of Saul? Choose a man for yourselves, and let him come down to me. [9] If he is able to fight with me and kill me, then we will be your servants. But if I prevail against him and kill him, then you shall be our servants

---

11 Jacob Bigelow, *Elements of Technology* (Boston: Hilliard, Gray, Little & Wilkins, 1829), *iv*.

and serve us." [10] And the Philistine said, "I defy the ranks of Israel this day. Give me a man, that we may fight together." [11] When Saul and all Israel heard these words of the Philistine, they were dismayed and greatly afraid.

Goliath was a giant, a champion and elite warrior, outfitted head to toe with the greatest weaponry he had plundered from across the ancient world. His battle tech was an assortment of superior pieces he amassed over the years as a professional warrior.

Saul was Israel's closest thing to a giant, head and shoulders taller than anyone else in the nation.[12] He was also their first king and the warrior most likely to be nudged out into this one-on-one fight. But he responded to Goliath in fear and unbelief. In Saul's place, a young shepherd stepped out in faith.

[32] David said to Saul, "Let no man's heart fail because of him. Your servant will go and fight with this Philistine." [33] And Saul said to David, "You are not able to go against this Philistine to fight with him, for you are but a youth, and he has been a man of war from his youth." [34] But David said to Saul, "Your servant used to keep sheep for his father. And when there came a lion, or a bear, and took a lamb from the flock, [35] I went after him and struck him and delivered it out of his mouth. And if he arose against me, I caught him by his beard and struck him and killed him. [36] Your servant has struck down both lions and bears, and this uncircumcised Philistine shall be like one of them, for he has defied the armies of the living God." [37] And David said, "The LORD who delivered me from the paw of the lion and from the

---

12 1 Sam. 9:2; 10:23.

paw of the bear will deliver me from the hand of this Philistine."
And Saul said to David, "Go, and the LORD be with you!"

Goliath had been killing men in battle for many years. He was
a pagan warrior bred to slay, an ancient terminator with metal-
shrouded flesh and tech-augmented strength. He was outfitted
with the latest body armor and engineered weapons, all supersized
to amplify his own native powers. In this story, as in other Old
Testament battles, God's people were poorly equipped to face off
against far more technologically powerful armies like the Philistines.

So when a young Jewish shepherd volunteered to fight the Philistine super-soldier, conventional wisdom said that David, too, must
be outfitted for war. So the boy tried on the king's war machinery.

[38] Then Saul clothed David with his armor. He put a helmet of
bronze on his head and clothed him with a coat of mail, [39] and
David strapped his sword over his armor. And he tried in vain to
go, for he had not tested them. Then David said to Saul, "I cannot go with these, for I have not tested them." So David put them
off. [40] Then he took his staff in his hand and chose five smooth
stones from the brook and put them in his shepherd's pouch.
His sling was in his hand, and he approached the Philistine.

The fundamental problem here is that David and Goliath were
mismatched in their energy potential. In ancient battles, the smaller
army was the underdog. Force wins battles, and the larger army
usually won. Whether we're talking about the combined kinetic
energy of sword-bearing field soldiers, the ferocity of horse-powered
chariots, the elastic potential energy behind arrows in full draw, the
explosive potential of gunpowder behind a bullet, or the energy

launched inside the warhead of a ballistic missile—wars are won by unleashing superior energy. In measurements of energy potential, Goliath was unrivaled, a weapon of mass dynamism, a small army in himself.

In a quick attempt to level the field and boost David's deficient power potential, Saul outfitted the young shepherd with his own war tech. David would have more power with armor and a sword, but verse 39 tells us that the boy had no experience with the gear. He lacked the proper technique. And without the technique, the advanced war tech was pointless because it could not do what it was made to do: amplify human energy and power.

Instead, David geared up with a familiar technique. Contrary to mistaken applications of this text that pit faith *against* technology, David had both. He had faith in God and technology at his side. David's whirling sling is a great example of technique—amplifying, focusing, and concentrating the animate energy of his arm to fire a smooth stone. David's sling was an early advance in the rich history of tech. That story began with levers and pulleys that amplified the power of animals and humans, then added more efficient inanimate power sources, like water (water wheels), wind (windmills), fire and coal (steam engines), electricity, fossil fuels, and nuclear power. The central storyline of human innovation follows how we discover more potent power sources, concentrate them, store them, and deploy them in demonstrations of greater and greater power.

So in this ancient one-on-one showdown we see a technological mismatch—but not in the direction we first assume. Goliath enters with technology suited for the front line in close-quarters combat with multiple enemies. David enters the standoff as a sniper. Assuming he has good aim, David proves to be the master technologist. His technology may be more primitive and useless in close-quarters combat. But as a projectile at this range, David's technique is superior. And yet his technique is small—small enough to put the focus on his faith. So David says to the giant: "You come to me with a sword and with a spear and with a javelin, but I come to you in the name of the LORD of hosts, the God of the armies of Israel, whom you have defied" (1 Sam. 17:45).

You know how it ends. The sling hits its mark. Goliath is knocked to the ground. David wields the giant's own sword and finishes the duel.[13] That sword will become David's sword.[14] And from this point, David will acclimate quickly to armor, shields, and blades.[15]

In the end, this epic face-off is not about whether technology is good or bad or whose technologies were better or worse. The point of the story is that in a clash between the gods of the Philistine giant and the living God of David, David's God wins. God's strength is made clear in David's weakness. That's the point. Whatever role human power and innovation play in this story, those roles are footnotes.

Yet we are left with a simple and profound example of two different levels of tech advance: cutting edge versus rudimentary. Both require technique. Both are technologies. Both amplify the power of their users.

The *Technium*

It's hard for us to appreciate the technology in this ancient battle because our powers today dwarf slingshots and swords. And old, animate power sources (like horses and oxen) are laughably weak

13 1 Sam. 17:50–51.
14 1 Sam. 21:7–9.
15 1 Sam. 18:4–5; 25:13.

in light of our modern, concentrated, inanimate power sources (like gas and electricity). We amass power into fuel tanks, batteries, and nuclear cores. But as I hope to show you in this book, all these advances are chapters in one big story.

Those chapters unfold like stages. First, technologies begin by amplifying and channeling animate power. Think of driving a carriage and using the leather lash to convert horsepower into the horizontal movement of wheels. Next come inanimate power sources under the direct control of humans. Think of driving the family minivan powered by exploding gasoline. These powers lead to a third stage, to semiautonomous systems that can operate apart from ongoing human input. Think of "self-driving" electric cars today, which still require the oversight of a human driver. The technologies we read about in the Bible all fall into the first stage. But our lives today are a mix of all three stages—spoons, cordless drills, and air-conditioners with thermostats.

Combined, our accumulated powers make us magicians. We can speed our bodies in a car at 70 mph. We can fly in an airplane at 575 mph. We can shoot a bullet at 1,700 mph. We can thumb a digital message to a thousand people at light speed. The power at our fingertips is truly remarkable.

But there's an emerging challenge on the horizon. Individual technologies that we can use are quickly becoming an ecosystem of technology we cannot escape. We have entered an age in which all of our techno-wonders are becoming so interconnected that they take on biological evolutionary characteristics—a seventh kingdom in nature, a unified and reinforced ecosystem. Kevin Kelly, cofounder of *Wired* magazine, calls the system the *technium*. Technology has reached a "self-amplifying" and "self-reinforcing system of creation," the point when "our system of tools and machines and ideas became so dense in feedback loops and complex interactions that it spawned a bit of independence."[16]

Inside this technium, older machines with various strengths get consolidated into new machines, with all their old powers added to even newer and more potent powers. "These combinations are like mating," writes Kelly. "They produce a hereditary tree of ancestral technologies. Just as in Darwinian evolution, tiny improvements are rewarded with more copies so that innovations spread steadily through the population. Older ideas merge and hatch idea-lings. Not only do technologies form ecosystems of cross-supported allies, but they also form evolutionary lines. The technium can really only be understood as a type of evolutionary life."[17] As a side note, many Christians find in Darwinism a sure explanation for the origins of biological life.[18] I don't.[19] But I also think that Kelly is right to use the theory of evolution as a metaphor for the tech age. Our machines mate by consolidating strengths. Supercomputers and robots inch their way toward autonomous intelligence, perhaps on a trajectory to a time when computers and robots will improve themselves without our help.

In evolutionary terms, every innovation of the future is built by recondensing or recombining the lineage of prior innovations into

---

16 Kelly, *What Technology Wants*, 11–12, 38.

17 Kelly, *What Technology Wants*, 45.

18 See Francis S. Collins, *The Language of God: A Scientist Presents Evidence for Belief* (New York: Free Press, 2007), 85–107. Collins claims evolution "as a mechanism, can be and must be true" (107). Without it "biology and medicine would be impossible to understand" (133). For a better take, see *Theistic Evolution: A Scientific, Philosophical, and Theological Critique*, ed. J. P. Moreland, Stephen C. Meyer, et al. (Wheaton, IL: Crossway, 2017).

19 See Michael J. Behe, *Darwin's Black Box: The Biochemical Challenge to Evolution* (New York: Free Press, 2006); *The Edge of Evolution: The Search for the Limits of Darwinism* (New York: Free Press, 2007); *Darwin Devolves: The New Science about DNA That Challenges Evolution* (San Francisco: HarperOne, 2019); and *A Mousetrap for Darwin* (Seattle: Discovery Institute, 2020).

new innovations. These first-gen innovations become ever-newer innovations in the future. Over time, they grow together into something of a unifying organism. In the end, writes Kelly, "this global-scale, circular, interconnected network of systems, subsystems, machines, pipes, roads, wires, conveyor belts, automobiles, servers and routers, codes, calculators, sensors, archives, activators, collective memory, and power generators—this whole grand contraption of interrelated and interdependent pieces forms a single system."[20] Very few technologies, if any, can be surgically extracted from this technium. So how do we respond?

The answer splits between dystopians and utopians.

On one hand, religious folks in particular tend to be tech dystopians and pessimists who view the technium as rebuilt Babel. Mankind is unified in rejection of God, in a technological evolution that God cannot stop, or chooses not to stop, until he eventually steps in and burns the whole experiment to the ground. The logical response for people of faith is to join the Amish outside the combustible city.

On the other hand, Darwinists and posthumanists tend to envision a world where human and machine blend together in a single existence, moving toward a heavenly utopia. They embrace the product of a technology that is "stitching together all the minds of the living, wrapping the planet in a vibrating cloak of electronic nerves, entire continents of machines conversing with one another."[21] The vision is a new and improved Babel 2.0, mankind reunified and augmented with innovation and machine power to self-exist forever.

I land somewhere in this mix, not a dystopian and not a utopian, but a Bible-believing creationist, Reformed in my theology,

20 Kelly, *What Technology Wants*, 8–9.
21 Kelly, *What Technology Wants*, 358.

trusting in God's providential orchestration over all things. I'm a city dweller concerned with the selfish motives at work in Silicon Valley, yet I'm also a tech optimist, eager to see and experience the future possibilities that lie ahead. In both cases, I'm sobered by a revelation that reminds me that the storyline of human tech will get fumbled and end badly too. I'll attempt to explain all this as we go.

The Path Ahead

This book is a roundtable with nine historic voices, framed by nine key texts of Scripture, as I seek to unseat twelve common myths about technology.

Here are brief profiles of the nine voices that will pop up throughout the book.

John Calvin (1509–1564), a French Reformer, celebrated theologian, and creationist who spawned an international movement that celebrated city building, culture making, and the scientific discoveries of non-Christians. He called Christians to hard work and frugality and put an end to "the religious and social stigma attached to wealth."[22] He brought peace between faith and science, opening the door for Christians to pursue science as an act of worship to God and love for neighbor.

Charles Haddon Spurgeon (1834–1892), a British pastor, Reformed Baptist, creationist, and one of the most famous Christian preachers in church history. A wide-eyed student of the cutting-edge innovations of his age, Spurgeon was Christ-centered and cut it straight about what technology could never accomplish.

22 Alister E. McGrath, *A Life of John Calvin: A Study in the Shaping of Western Culture* (Hoboken, NJ: Wiley-Blackwell, 1993), 219–61.

Abraham Kuyper (1837–1920), a Dutch neo-Calvinist, theologian, journalist, and one-time prime minister of the Netherlands. Kuyper was a creationist who took Calvin's worldview, pushed it to its optimistic limit, and celebrated the common grace of man's scientific future.

Herman Bavinck (1854–1921), a Dutch neo-Calvinist, widely celebrated theologian, and creationist who built from Calvin's vision a cautious approach toward innovation. Bavinck identified the spiritual challenges of the technologies of the past, the present, and the future.

Jacques Ellul (1912–1994), a French philosopher, Christian, and tech pessimist who believed that every innovation introduces more problems than solutions. Ellul protested against the economic and political technocracy that stood in direct conflict with Christian discipleship.

Wendell Berry (1934–), an American novelist, essayist, and conservationist known for his advocacy of rural life and belligerence toward big tech. Berry frames his conservationism through a Christian worldview, albeit in ways that are a little shallow on doctrine.

Kevin Kelly (1952–), an American cofounder of *Wired* magazine, conservationist, and decades-long reporter from the front lines of American technology. Kelly is a tech optimist in vision but a tech minimalist in application—a lifestyle he adapted from the Amish. He's a Darwinist, claims a religious conversion experience, and reconciles God and tech via open theism, the idea that God watches with surprise to see what we will invent next.

Elon Musk (1971–), an American billionaire, eccentric entrepreneur, and technologist behind some of America's most ambitious companies like Tesla, SpaceX, and Neuralink. He is pushing forward space exploration with the goal of colonizing Mars but is known more immediately for his successful endeavors in electricity and self-driving cars. When asked if science and religion can coexist, he said, "Probably not."[23] Musk advocates simulation theory, that we don't live inside a base reality but exist more likely inside one of many Matrix-like simulation programs designed by a superior intelligence.

Yuval Noah Harari (1976–), an Israeli professor of history, adamant atheist, and bestselling author who earned the title "The Historian of the Future." A convinced Darwinist, Harari is an Orwellian tech dystopian attempting to shake people with two predictions in the form of two new religions: *techno-humanism*, a world of genetically modified superhumans augmented with new computing powers; and *dataism*, where ultimate authority rests in the most powerful computing being, once man, soon to become artificial intelligence (AI).

Alongside a conversation with these nine voices (and a few others), the book is organized around the study of nine key sections of Scripture: Genesis 4:1–26; 6:11–22; 11:1–9; 1 Samuel 17:1–58; Job 28:1–28; Psalm 20:1–9; Isaiah 28:23–29; 54:16–17; and

---

23 SoulPancake, "Elon Musk Captured by Rainn Wilson!" youtube.com (Mar. 18, 2013).

Revelation 18:1–24. Many others could be added, but these are the most important.

As we study these important blocks of Scripture, may I ask a favor? As readers, we tend to skim indented quotes (I know, because I do it too). But please don't. Please read every indented text with special care.

As we move along, I'll highlight key takeaways and dispel the most common myths about technology I hear and see in the church, particularly these twelve:

Myth 1: Human innovation is an inorganic imposition forced onto the created order.

Myth 2: Humans set the technological limits and possibilities over creation.

Myth 3: Human innovation is autonomous, unlimited, and unchecked.

Myth 4: God is unrelated to the improvements of human innovation.

Myth 5: Non-Christian inventors cannot fulfill the will of God.

Myth 6: God will send the most beneficial innovations through Christians.

Myth 7: Humans can unleash techno-powers beyond the control of God.

Myth 8: Innovations are good as long as they are pragmatically useful.

Myth 9: God governs only virtuous technologies.

Myth 10: God didn't have the iPhone in mind when he created the world.

Myth 11: Our discovery of atomic power was a mistake that God never intended.

Myth 12: Christian flourishing hinges on my adoption or rejection of the technium.

Faith and Physics

Since before the Enlightenment, science and the church have often been friends and sometimes enemies. The tension was not always the fault of science. This clash is unfortunate, because in a grassy valley in the middle of ancient Israel, God's man, David, wielded physics and faith at the same time. Can we learn to do the same? Can we find a life of faith within this world of amplified human possibility? Can we find a place where God-centered trust and technique-wielding skill complement one another?

The agnostic technocrat thinks that he must shove God aside for technology to flourish. The Christian agrarian thinks that he must shove technology aside in order for faith to thrive. But both the tech optimist and the tech pessimist sell God short. Even the most procreation, material-celebrating forms of Christianity struggle to know what to do with smartphones, space exploration, and gene-based medicine.

Christians rebuke gnosticism. In Christ we celebrate the material world, like freshly brewed coffee, blossoming fruit trees, hot bread, soft butter, and warm honey. Nature and gardens and sunshine and play and laughter are gifts to be enjoyed. So, too, are dances, weddings, and married sex. But should we also celebrate the smartphone, the microprocessor, and the nuclear core? If it plugs into the electrical grid, can we celebrate it?

People of faith have sometimes undermined thoughtful conversations on technology by dismissing human innovation with terms of domination (like *technopoly*) and a few other -isms (like *technicism, scientism,* and *economicism*).

I think we need a new discussion, and this book is my attempt. My previous book covered Christian living in the attention economy.[24] Before that, I wrote a book on smartphones and how digital technology is changing our lives. There I first laid out a brief, ten-page outline of how I understand the tech world through Scripture.[25] Over the following years, that summary generated robust conversations, and I knew that I would need to develop my outline into a book. So here it is, my -ology of technology, my biblical theology of technology.

One of my original titles for this book was *A Christian Optimist's Guide to Modern Technology*. Tech is not all roses, but it's not all bad Apples either. This book is my case for a more positive view of human innovation and innovators. As a tech optimist, I know that this book would market better as an alarmist, doomsday warning about how Satan hijacked the electrical grid, controls us through our smartphones, and wants to implant us with the digital mark of the beast. I would sell you a vast conspiracy coupled with a theology of a powerless god who doesn't know what to do. I would put the future of the world in your hands as our only hope. I would focus your attention on the scariest new tech so you would ignore the glories of the vast tech advances that adorn your daily life. I would end with an appendix on how to dig out a bunker for a rural, off-grid commune. And I'd write the whole book with the caps-lock on. Fear sells books, but my theology—what I know about the gloriously sovereign Creator and his incredible creation—forbids me from stoking more fear. So I'm optimistic—not optimistic in man, but in the God who governs every square inch of Silicon Valley.

---

24 Tony Reinke, *Competing Spectacles: Treasuring Christ in the Media Age* (Wheaton, IL: Crossway, 2019).

25 Tony Reinke, *12 Ways Your Phone Is Changing You* (Wheaton, IL: Crossway, 2017), 29–39.

In the pages ahead, I extend my research beyond media and smartphones to find answers that have alluded the world, from Babel's tower to SpaceX's rockets. "Technology in fact is one of the most completely known parts of the human experience," writes technology theorist Brian Arthur. "Yet of its essence—the deep nature of its being—we know little."[26] This is true both outside and inside the church. Do our innovations threaten God? Do they make him more irrelevant to life? What is God's relationship to Silicon Valley and Silicon Alley? How does he relate to our most impressive innovators? Is God threatened by the technium? Where do our technologies and gadgets come from? What can technologies do for us? What can they *never* do for us? And how much tech is too much tech in the Christian life?

We need answers.

26 W. Brian Arthur, *The Nature of Technology: What It Is and How It Evolves* (New York: Penguin, 2009), 13.

2

What Is God's Relationship
to Technology?

WHEN I THINK ABOUT TECHNOLOGY, I picture beeping alerts and blinking LED lights on a small gadget. I don't typically think about tar. But that's where our journey starts—with sticky, black tar. It will help us answer one of the core questions we face as Christians in this age: What is God's relationship to human technology?

How you answer this question will frame your own relationship with technology. It will affect how you decide what innovations to adopt, what industries to work in, and what guidelines to set for the tech you use. It's a huge question, and it all goes back to tar.

Our journey into the black stuff begins in Genesis 6:11. Sin has entered into creation at this point. Sin has disrupted *everything* and corrupted *everyone*. Watch for the tar.

[11] Now the earth was corrupt in God's sight, and the earth was filled with violence. [12] And God saw the earth, and behold, it was corrupt, for all flesh had corrupted their way on the earth.

[13] And God said to Noah, "I have determined to make an end of all flesh, for the earth is filled with violence through them. Behold, I will destroy them with the earth. [14] Make yourself an ark of gopher wood. Make rooms in the ark, and cover it inside and out with pitch."

There it is: *pitch*, or *tar*.

Global rebellion calls for global response. God tells Noah that he is going to destroy everything on earth. Then he gives Noah the blueprints for a massive building project that will call on every known feat of human engineering and technology to build an ocean-worthy ship. Within a society that used only small boats to navigate rivers, the ark defied all existing categories of engineering.[1] By sheer size, it exceeded all practical use. And by sheer necessity, it forced Noah to begin blazing new paths in engineering and technology in a building project that would require over a century to complete.

As a final step before the storm clouds gathered, Noah gathered tar. Where did he find it? Perhaps, even prior to the flood, coal and oil and tar were naturally available in creation.[2] Or maybe Noah distilled pitch from pine tree resin.[3] Either way, Noah slathered his ark with thousands of gallons of tar, inside and out, to keep the hull of the ship waterproof in the face of God's coming global judgment.

A Long Shadow of Hideous Strength

Forty days and nights later, the ark touched down and humanity quickly filled out into a comprehensive table of nations detailed in

1   Abraham Kuyper, *Common Grace: God's Gifts for a Fallen World* (Bellingham, WA: Lexham Press, 2020), 1:338.
2   John Matthews, "The Origin of Oil—A Creationist Answer," answersingenesis.org (Dec. 17, 2008).
3   Tas B. Walker, "The Pitch for Noah's Ark," creation.com (July 20, 2016).

Genesis 10. Due to extraordinarily long life spans at the time, just 150 years after the flood the population could have easily climbed to fifty thousand. But instead of spreading across the globe, this postflood population gathered together in Babel, the second major story of tar and the second major story of technological innovation in the early chapters of Genesis. Here's what we read in Genesis 11.

[1] Now the whole earth had one language and the same words.

The globe's fifty thousand residents had a single line of cultural continuity that ran back to Noah. They had one language, the same words. And they started to make a new home on earth.

[2] And as people migrated from the east, they found a plain in the land of Shinar and settled there. [3] And they said to one another, "Come, let us make bricks, and burn them thoroughly." And they had brick for stone, and bitumen for mortar.

There it is again—*bitumen*, or *tar*, the same concept we saw in the ark.[4] The Babelites discovered tar, possibly naturally occurring tar. More noteworthy, they also invented a new way to fire bricks. The fire-hardened brick is novel tech leading to new possibilities. "The citizens did not start out deciding to build a city and tower," theologian Alastair Roberts points out. "They start off by discovering a new technology, fired bricks, and having discovered the technology, they decide to build a tower and city. There is

---

4 Noah was commanded to use pitch, a different word, on his ark. But Moses's ark is made buoyant with bitumen and pitch, linking the two terms. "When she could hide him no longer, she took for him a basket made of bulrushes and daubed it with bitumen and pitch. She put the child in it and placed it among the reeds by the river bank" (Ex. 2:3).

something about technologies that open our imagination to new possibilities, preceding our sense of what can or should be done."[5]

Out of the novel tech (fired bricks) and the natural discovery (tar) come new urban dreams of what is now possible.

> [4] Then they said, "Come, let us build ourselves a city and a tower with its top in the heavens, and let us make a name for ourselves, lest we be dispersed over the face of the whole earth." [5] And the LORD came down to see the city and the tower, which the children of man had built.

With these new technological possibilities, man's first global vernacular breaks forth in aspiration. "Let us build." "Let us make." Some primitive impulse inside us demands to innovate. So in unison, humanity struts forward to collaborate on a building project of autonomy. This is not a tower for mere reputation; it is a tower of independence, the desire "to be definitively separated from God."[6] To reject God's commission to spread across the globe, man engineers a tower into the sky.

Before long, God notices the rebellion, descends from heaven, bends down to his knees, puts a cheek on the ground, and squints to see this tower and measure its height. (This is divine sarcasm.) And then he responds.

> [6] And the LORD said, "Behold, they are one people, and they have all one language, and this is only the beginning of what they will do. And nothing that they propose to do will now be impossible for them. [7] Come, let us go down and there confuse their language,

5 Alastair Roberts, phone conversation with the author, October 29, 2020.
6 Jacques Ellul, *The Meaning of the City* (Eugene, OR: Wipf & Stock, 2011), 16.

so that they may not understand one another's speech." [8] So the LORD dispersed them from there over the face of all the earth, and they left off building the city. [9] Therefore its name was called Babel, because there the LORD confused the language of all the earth. And from there the LORD dispersed them over the face of all the earth.

Babel is a city and tower complex. The city-tower is mentioned twice, and then finally just the city. Some think that's because God wants to make the point that he doesn't want the city completed. Perhaps. Or maybe the tower was complete and the city was in progress. We don't know. We do know that it was a city-tower and that it was a premature attempt at utopia. God called the first family, Adam and Eve, to spread out and fill the earth. And after the flood he called Noah's family to do the same. And then mankind collected itself into one herd and settled into one unified city-tower. But that was not God's plan. The result was a city-temple to human rebellion—a religious center, not an urban center of atheists and agnostics. Babel was the new global epicenter of human worship. All of humanity gathered together, with religious intent, with what appeared to be the goal of opening a portal in the sky, storming heaven, dethroning God, and enthroning humanity in his place.

And yet for all its aspiration, Babel, like any human engineering feat, poses no threat to God. Verse 6 highlights Babel's problem: not that it makes God insecure but that it sets man on a new path of self-confining self-destruction. Man's increasing ambition and power don't threaten God; they threaten man himself, because "the more power they are able to concentrate, the more harm they will be able to do to themselves and the world."[7]

7   Donald E. Gowan, *From Eden to Babel: A Commentary on the Book of Genesis 1–11*, International Theological Commentary (Grand Rapids, MI: Eerdmans, 1988), 119.

Tech in Babel

The Babelites' attempt at a tower was likely a ziggurat, a glorified pile of LEGOs, built from baked bricks, not from uncut stones. Their bricks were novel tech. Fire-baked bricks are less likely to flake apart or crack. They hold up well, especially under the harshest weather conditions. They are stable, strong, and meant to endure for ages. And the baked bricks were pasted together with black tar—bitumen. Today we appreciate the wealth of oil deposits in the Middle East. But long before oil rigs and drills ignited the petroleum industry and brought untold wealth to the inheritors of Babel's lands, that same land oozed crude oil on the surface.[8] It was only a matter of time until the depths of these deposits would be discovered, drilled, siphoned, and poured into barrels to power the modern world with a wealth and power that Babel's original builders could have never imagined. The story of Middle Eastern oil begins here, in Babel's tar.

But scholars propose that this tar was an engineering fail, a mismatch, like building a skyscraper of steel beams but connecting those beams with duct tape instead of bolts. On one hand, the baked bricks symbolize "permanence and stability." But the tar, we are told, is "hardly an adequate mortar," so inferior that we are seeing here a "flaw in the blueprint," says one theologian.[9]

On the contrary, I think the design was brilliantly calculated. Historically, only one and a half centuries separated the flood from Babel. So the people of Babel remembered Noah's ark. The flood was public knowledge, never drifting very far from the collective memory. This means two things.

---

8   W. Sibley Towner, *Genesis*, ed. Patrick D. Miller and David L. Bartlett, Westminster Bible Companion (Louisville, KY: Westminster John Knox, 2001), 109–10.

9   Leland Ryken et al., *Dictionary of Biblical Imagery* (Downers Grove, IL: InterVarsity Press, 2000), 66.

First, the only flood survivors were Noah, his family, two animals of every kind, and the ark itself—the most incredible technological achievement in the history of mankind up until that point, a compilation of all the building technologies prior to the flood—carried over and preserved into a new age. The ark helped to inspire the technological advances that would lead to the aspirational ziggurat of Babel. It carried more engineering feats than just ship building (as we will soon see). But the colossal ark alone must have filled humanity with engineering aspiration.

Second, knowledge of the flood meant that the postflood people all knew that God judged sinful humans who spurned him and rebelled against his will. Everyone knew it. So if you're going to build a tower to dethrone God and break free from him, you'd better be ready. Ready for what? Ready for judgment. Ready for God's rage to once again pour down from the skies in catastrophic flooding. So you harden your bricks with fire, and you glue them with tar. You make your tower watertight. And only then do you stand on the top of the roof, look up into the blue heavens, raise your fist, and say: "Good luck washing us away now!"

Expelled (for Good)

All of this human engineering was meant to thwart God, or so the Babelites thought. Of course, God did judge them. He dethroned them by confusing their single language and dispersing them across the globe. Which is another odd part of the story, because it doesn't really matter what you call a baked block or a bucket of tar. Call it whatever you want; you can continue building. "Bring me that." "Put it here." "Glue it with this." You don't need the shared vocabulary of Shakespeare to finish off a half-built city. You can communicate with pointed fingers. So it's not as if God caused everyone to use different nomenclature for *tar* and *block* and the work screeched to a halt. Something more was happening here.

God sized up the tower and scattered humans across the globe. This judgment brought two seismic changes. First, by multiplying languages and spreading mankind across the globe, God introduced new global networks and technologies of communication, travel, and shipping. These industries were now inevitable.[10] Second, by multiplying languages and spreading mankind across the globe, he introduced new global tensions among humans that would help protect us from our future technological innovations. Let me explain.

If you love learning languages, you know that language is more than a dictionary. Every language has an inner logic that connects to the culture. Whole worldviews are reflected in a dialect. Every culture produces its own musical sounds and its own shape of bread. Differing languages represent differing cultures and unique ways of designing cities, towers, and homes. "Many anthropologists say that if you truly want to learn about a people group, you need only to learn their language and it will tell you all you need to know. So by confusing their languages, God was essentially reprogramming their sense of self, their relational connections, and how they viewed the world."[11] Simultaneous to the plethora of new languages, we see evidence that God introduced hundreds of new ways to think about the world. Those new thoughts produced a multiplication of new religions. When people aiming to overthrow God were scattered, they produced competing idols and

10 Kuyper, *Common Grace*, 2:588.
11 John Dyer, *From the Garden to the City: The Redeeming and Corrupting Power of Technology* (Grand Rapids, MI: Kregel, 2011), 105.

tribal deities in their own image. Babel even marks the genesis of ethnic animosity.[12]

But here's the big point relative to technology. In the multiplication of languages in Babel, "God keeps man from forming a truth valid for all men. Henceforth, man's truth will only be partial and contested."[13] And this worldview disharmony trickles down into how a given culture creates and uses (or rejects) certain innovations. So in Babel, instead of one unified way to design a tower and a city, there became a hundred opinions on how to do it best. For a long time, theologians have said that this multiplication of languages is the origin of all cultures in the world. What we must add to the discussion is a point about technology. By multiplying cultures, God coded into the drama of humanity different ways of thinking about and engaging with the world. These differences are so potent that they will help restrain us from adopting any one, single technology.

Babel was no accident. Babel was the product of man's aspiration and innovation, which God intended to hack all along in order to create all the cultures of the earth and to set in place a global subversion that would fracture human aspiration and undermine universal tech adoption in the future.

The Good News of Disunity

Beginning in Babel, universal consensus was made impossible. And that's good news. Here's a modern example why. In the fall of 2019, Delta Airlines began using facial recognition in the United States

---

12 Bavinck calls Babel the beginning of "race instinct, sense of nationality, enmity, and hatred," the spring of all "divisive forces between peoples," "an astonishing punishment and a terrible judgment" never to be undone by human programs or culture, only by Christ and within his Bride on earth. Herman Bavinck, *The Wonderful Works of God* (Glenside, PA: Westminster Seminary Press, 2019), 35.

13 Ellul, *Meaning of the City*, 19.

at dozens of domestic gates and for all international flights. This tactic was possible because our biometric data, the unique vector of each person's facial features, is held in a government database, which Delta gained access to. Biometric boarding—using a facial scan, not a boarding pass—is faster, easier, and, according to Delta, more popular with customers.[14] Facial recognition is widely adopted in America. Why? Personal convenience.

Simultaneously, on the other side of the globe, masked protestors in Hong Kong wielding cordless Sawzalls chopped down lamp poles believed to carry the government's facial recognition cameras, to protest a lack of human rights. Likely no country in the world has a larger database on its citizens than China, a country long suspected for using private data to feed artificial intelligence (AI)–driven coercion. Facial recognition is widely protested in China. Why? Personal peril.

I'm not suggesting that your opinion on the governmental use of biometric data depends on whether you speak English or Cantonese. This one illustration shows how multiplied cultures introduced resistance into the adoption of any one technology. Instead of allowing humanity to live within a global consensus (as in Babel), God broke in, confused the languages, and multiplied cultures. He coded internal tensions and disharmony into the drama of humanity, tensions that will help check and limit the adoption of technologies in a fallen world. Eager adoption of biometric data is held in check by fears of state coercion.

So God's judgment in Babel introduced a hundred new and competing opinions about how to best build a city. In fact, this is what researchers find when they look at how technology distributes

---

14   Kathryn Steele, "Delta Expands Optional Facial Recognition Boarding to New Airports, More Customers," delta.com (Dec. 8, 2019).

throughout the globe today. Undeniable ethnic biases still determine which advances get adopted or rejected in a given culture.[15] *Now* add the vernacular confusion and you get a sense of the scaled frustration that brought an abrupt end to Project Babel.

By squelching human consensus and spawning a biodiversity of cultures, God created for the first time an inherent tension within humanity. We see it in spiritualists and native tribespeople of New Zealand who are on the front lines to preserve the country's unique biodiversity by resisting foreign genetic modifications being pushed into the ecosystem by scientists.[16] We see it in the rise of Elon Musk–type characters and the tech-pioneer class of Silicon Valley. And concurrently, we see it in the rise of Wendell Berry and tech-resistant communities like the Amish. For Berry, the world went to the dogs as soon as man stopped limiting himself to the powers available in animate sources—but could stockpile power in batteries, fuel tanks, and nuclear cores. For Musk, his whole enterprise is built on the amassing of electricity in giant lithium battery farms and in SpaceX rockets propelled by tons of dense, rocket-grade kerosene and methane. Mastering technology, according to Musk, is aspiring to make more of it. Mastering technology, according to Berry, is learning to ignore most of it.

The aspirational mind envisioning new tech must learn to live alongside the mind that foresees and warns of big tech's tyranny. *Only* listen to the ambitions of Musk, or *only* read the cautions of Berry, and I think you will lose a grip on the broader reality. We are healthier for the cultural tension that exists between them, and this tension began in Babel. From the scattering in Babel until the

15 Kevin Kelly, *What Technology Wants* (New York: Penguin, 2011), 291.

16 A dominant theme in the documentary *Unnatural Selection*, produced by Radley Studios et al., distributed by Netflix (2019).

ultimate defeat of Babylon, human innovation can only unfold within this unalleviated tension. And this is a mercy from the Lord, especially as pressures mount to embrace genetic modifications and AI. But this is a temporary mercy (as we will see later).

For now, I ask: What is God's relationship to human innovation and technology? In Noah, he commanded it. In the ark, God took human engineering and technology and wrote it into the grand story of redemption. But in Babel, God squashed it. In the face of human self-glory, he introduced the tensions that utterly thwarted human collaboration.

Is God Threatened by Babel?

Now, if we stop with the words of Genesis 11:6, and if we miss the sarcasm in the story, and if we miss the dangers that the tower represents (not for God, but for man's own self-destruction), we may be left to assume that God's relationship to human innovation is analogous to Homer Simpson at the control panel of a nuclear power plant during a meltdown, scuttling around with frightened yelps, unable to fix anything by his attempts to hit random buttons and pull random levers, unaware of any sequence of frantic actions that will stop the impending meltdown.

If we end with Genesis 11:6, God seems caught off-guard by human innovation. He seems surprised and aloof over what emerged in Babel. He appears to wield the power only to *respond* to human innovation—to squelch it, to snuff it out. He even appears defenseless by all the future possibilities of human innovation. And if we're being honest, many Christians operate with this assumption. In the face of human possibility, God seems aloof, surprised, alarmed, even threatened. But such a conclusion is very wrong, as the prophet Isaiah shows us.

Blacksmiths, Fire, and Swords

Any misunderstandings with Genesis 11:6 get cleared away by our next important text: Isaiah 54:16–17. Up to this point, we've only talked about technology horizontally. Swords and slings and the ark and Babel's city-tower are simply the *products* of engineering technology. We have yet to answer the question, Where do innovators come from?

God answers this question directly in Isaiah 54:16–17, as he speaks comfort to his people.

> [16] Behold, I have created [*bara*] the smith
> who blows the fire of coals
> and produces a weapon for its purpose.
> I have also created [*bara*] the ravager to destroy;
> [and yet] [17] no weapon that is fashioned against you [God's
> people] shall succeed,
> and you shall refute every tongue that rises against you in
> judgment.
> This is the heritage of the servants of the LORD
> and their vindication from me, declares the LORD.

Illumined by the promise of a new covenant, having forecasted the death and resurrection of his servant in the chapter before, God claims supremacy over armies. Isaiah has made this point earlier.[17] But here, God makes three incredibly specific claims: (1) He creates the creators of weapons. (2) He creates the wielders of those weapons. (3) He governs the outcomes of those weaponized warriors—the ravagers.

17 Isa. 10:5–34; 13:1–22.

45

The Smith

Let's start with the ancient blacksmith. As the elite tech class of the ancient world, blacksmiths led the charge in human innovation for centuries, especially in the age of the Old Testament.

Ancient blacksmiths possessed the deep secrets of a magic called metalworking, learned techniques confidentially passed from one generation to another by years of training. Across ancient cultures, blacksmiths lived together in guilds in order to "jealously guard their secrets and adhere to a rigid system of ethics." Their work was suffused with tradition and ceremony and "rites of purification, fasting, meditation, prayers, sacrifices, and other acts of worship." They kept rituals over the smelting and forging processes, and animal sacrifices kept their holy forge fires pure. Blacksmiths decontaminated small bits of the world by fire and created holy objects with new spiritual force. "In folklore, iron objects are traditionally protective against witchcraft, evil spirits, and malign influences. The power of the metal is often ascribed to its connection with the earth, that is, it is believed to be a piece of earth that has been purified by fire." Even the tools of the smith were handled with holy reverence. The smith's entire forge—his hammer, anvil, and furnace—became a "ritual center," a temple where the smith took his cultural place as a priest over creation, enacting his holy duties to purify the earth.[18]

Copper and bronze gave way to iron tools rather slowly because iron ore smelting requires very high temperatures of charcoal and bellows, which arrived later in the history of metallurgy.[19] Isaiah's sword maker, like all the earliest iron smiths, more com-

18 Paula M. McNutt, *The Forging of Israel: Iron Technology, Symbolism and Tradition in Ancient Society* (Decatur, GA: Almond Press, 1990), 45–46.
19 Eugene H. Merrill, *Deuteronomy*, vol. 4, New American Commentary (Nashville, TN: B&H, 1994), 186.

monly worked with ancient meteors that had impacted the earth's surface, adding to the mystique of their labors. "Primitive man everywhere used meteoric iron in the earliest stage of his metal culture," writes one anthropologist. "That is to say, when he was beginning to use the native metals—gold and copper chiefly—such as he found ready to his hand on the surface of the ground. Iron rarely occurs native, but it is obtained in the shape of meteorites, the dropping of which from the sky provided man with a metal of remarkable excellence." Meteoric iron was excellent because it contained nickel. Ancient people called this iron "fire from heaven" or "metal from heaven."[20] The meteor's heavenly origin made the priestly role of the blacksmith even clearer. He was a mediator standing between the iron of heaven and the people of earth, a whole industry standing between the divine realm and the earthly realm.[21]

Blacksmiths forged tools of war, tools of commerce, and tools to ward off spirits. They fashioned heaven's gifts with spiritual techniques and became a power class of innovators. Their inventions amplified human power and made the warrior and the nation's army more fierce and deadly. Smiths stood at the center of the ancient tech industry. They were saviors of sorts, masters of iron, makers of ancient warfare technology, creators of power and safety.

The smith was the chief innovator of his age, an expert maker of swords and spears and shields and axes. And yet within this age of hardened iron, of war machines hardened by flame, God reigned supreme over every part of it. The blacksmith creates powerful swords, yes. But God creates blacksmiths. God introduced the

---

20 T. A. Rickard, "The Use of Meteoric Iron," *Royal Anthropological Institute of Great Britain and Ireland* (1941), 71:55–65.
21 McNutt, *Forging of Israel*, 264.

skill to work metal into human history. God sanctioned the forge and the anvil. God mixed the meteor's metals and threw them to earth. The entire metalsmithing guild owes its origin to God as it serves his providence.

The Ravager

God's sovereignty doesn't stop with the innovators, with smiths who hammer and sharpen weapons. His omnipotence carries over to those who *wield* those weapons too. The same dynamic is at work in Isaiah 54:16, when God said, "I have also created the ravager to destroy." The word *ravager* is a summary of all the effects that you would imagine from someone wielding a sword, spear, or axe destructively—ruining, breaking, pillaging, seizing, laying waste.

God *makes* every sword wielder, even the ravager—the destroyer. But he makes them not in some vague or generic sense of permitting them to be. The repeated Hebrew in this text for "created" (*bara*) is a very specific word reserved for God's work of creation as the sole originator, here the sole originator of the smith and the ravager. *Bara* marks the "historical continuation" of God's creative activity into the present.[22] The Creator still creates today, as he created at the beginning of Genesis, but in narrower and more particular ways. And yet *bara* still "contains the idea both of complete effortlessness and *creatio ex nihilo*, since it is never connected with any statement of the material."[23] In order to *bara*, God needs no raw ingredients. He alone effortlessly makes from scratch what is not inevitable. That is his relationship to the smith

---

22 Hans-Helmut Esser, "Κτίσις," in *New International Dictionary of New Testament Theology* (Grand Rapids, MI: Zondervan, 1986), 379.

23 Gerhard von Rad, *Genesis: A Commentary* (Louisville, KY: Westminster John Knox, 1973), 47.

and the ravager. Apart from him they would never exist. Only by his will and design, they do.

The sword-making smith and the sword-wielding ravager are both ordained by God in his sovereign decree. God is present over the smith and the ravager as their sole maker. The ravager himself is forged by God alone, and he is forged for the purpose of plundering. God governs each creature toward good ends directly; and he governs over all sin and evil indirectly. He governs the lives and decisions of all his creatures to his wise end, a sovereign truth commonly called the doctrine of concurrence. God is the *primary*, but *remote*, cause of all human action. Humans are the *secondary*, but *proximate* (or *near*), causes of their own actions.[24] Thus, even the scope of the ravager's destruction is sanctioned by God. He governs human technology and how those technologies are wielded, even destructive ones. He does all this by secondary causality.

It is sobering but true that "nothing occurs, not even the destroying acts of the enemies of God's people, apart from God himself."[25] This is a heavy but essential text in the Bible (and there are many of them), in which you begin to understand that God not only governs gentle butterflies, fluffy puppies, healing technologies, and life-saving medications. He wields a greater, more comprehensive governance over everything. God brings "well-being" and he "*create*[s] calamity" (Isa. 45:7). "Create" here is again the Hebrew word for God as the sole origin. Only God can create well-being and create calamity, because only God has exalted himself to a position of utter transcendence.[26]

---

24 Scott Christensen, *What about Free Will? Reconciling Our Choices with God's Sovereignty* (Phillipsburg, NJ: P&R, 2016), 77–81, 254.

25 Edward Young, *The Book of Isaiah, Chapters 40–66* (Grand Rapids, MI: Eerdmans, 1972), 372.

26 Scott Christensen, *What about Evil? A Defense of God's Sovereign Glory* (Phillipsburg, NJ: P&R, 2020), 186–89.

No one is like our God. Blessed be the name of the Lord.[27] He has providential intent for termites, rattlesnakes, social chaos, and even destroyers who wield the latest tech advances for destruction.

All of these points are made so that God can reassure his people: "No weapon that is fashioned against you shall succeed" (Isa. 54:17). Why not? Why can God make such a huge claim over the security of his people? Because of the verse before: "I . . . created the ravager to destroy" (Isa. 54:16). To carry out his own secret purposes, God creates and governs over every ravager in this world. The world's greatest threats, even at their most technologically outfitted, can only wield a power *given* and a purpose *governed* entirely by God.

Now, my mind naturally imagines a safer world for the church if ravagers never existed in the first place. But that's not how this text works. Instead, the reasoning goes like this: because God creates and wields the destroyers inside this fallen world, his people should be confident of his protective mercies. God governs every human innovation, even destroying ones. From the universe itself, to shalom and chaos, from every smith and ravager, to every maker and wielder of war tech—this is all God's work. By his creative work alone, and by his divine orchestration over all his creation, he rules over humanity in every way. If God makes each weapon by creating each weapon maker, and then creates each wielder of each weapon, "we should not think that anything can come to us that will contradict God's purposes for us."[28] That's divine logic, and very different from how we naturally reason things out.[29]

27 Job 1:21.
28 John N. Oswalt, *The Book of Isaiah, Chapters 40–66*, New International Commentary on the Old Testament (Grand Rapids, MI: Eerdmans, 1998), 430.
29 Isa. 55:8–9.

In ways we find hard to understand, ravagers are created to carry out God's will. As I write this chapter, parts of Minneapolis burn from riots. Each night I watch with horror as hundreds of stores are looted and local businesses are overrun by rioters and burned to the ground. This on-screen *ravaging* hits very close to home (just a few blocks from my previous home). And this ravaging is wicked. So I don't assume that Isaiah 54:16–17 is simple or easy to embrace. It's not. And I'm sure that some of you have wrestled with it or are wrestling with it—that God raises up weaponized ravagers to wield his judgment on earth. And it's okay to wrestle with this. I cannot rush you along. But I do think that the prophet Isaiah is clear, as are other writers of Scripture who show how God wields ravagers.[30] God is unchallenged and unmatched in his supremacy. He creates light. He creates darkness. He creates shalom. He creates calamity.[31] He reigns over everything, even ravenous warfare. He has a secret will, and that secret will raises up and wields big tech as he sees fit.

The ravager will never stop the church on earth. And yet God's sovereignty over the ravager should never be taken to mean that Christians are impervious to all harm. That is not my claim. Everything is amplified in the technological age, especially the self-damaging fallout of our hubristic mishandling of power. All our tech making should be sobered by a healthy fear of the Aberfan disaster, when a pile of mine waste became a waterlogged mudslide that killed 116 children and twenty-eight adults in Wales. We should be sobered by a healthy fear of the Bhopal disaster in India, a pesticide-manufacturing gas leak that killed thousands and infected a half-million people. We should fear the damages that potent

30 See also Jer. 22:6–9; 51:20–23; Ezek. 9:1–11; 21:28–32.
31 Isa. 45:5–7.

medications can bring, like when thalidomide was prescribed to pregnant women to offset pregnancy pains and discomforts but led to tens of thousands of babies born with unspeakable physical deformities. And we should never forget Chernobyl, the nuclear disaster that permanently evacuated and poisoned a city, all started by a human test gone wrong.

These are real tragedies that hurt real people and call for real tears and demand real reform in our practices. But in all these disasters, we are wrong to assume that God was absent. In reality, God is sovereignly present even when our technologies destroy. He is present and good and bringing about—by every intentional evil of a ravager or by every accidental evil from a technological disaster—a million resulting consequences in countless lives, according to his good and wise purposes that we cannot immediately see.[32]

So God created ravagers. They do real harm. They cannot destroy God's people. But they are appointed for a providential end, for a good purpose that could otherwise not come through any other means. And each ravager will be judged for his evil actions. Thus, says Calvin, "we must not, on that account, lay blame on God, as if he were the author of the unjust cruelty which dwells in men alone; for God does not give assent to their wicked inclinations, but regulates their efforts by his secret providence" so that he can employ them, when needed, as his "instruments of anger."[33] Within God's good plan of providence is a place for sword-wielding ravagers who level ancient cities. No matter how nefarious their motives, the smith and the ravager exist by divine

32 À la Gen. 50:20.
33 John Calvin, *Commentary on the Book of the Prophet Isaiah*, trans. William Pringle (Edinburgh: Calvin Translation Society, 1853), 4:152. See also the Westminster Confession of Faith, 3.1–8.

appointment because their activity in this world ultimately brings about a greater good in God's greater glory than if they did not exist.

Here's my point. In any discussion of technology, many Christians get hung up on the most powerful technologists in the world who are inventing the most threatening innovations on earth—nuclear power, killing weapons, space rockets, modified genetics—and assume that these men and women fall outside God's governance. They don't. Isaiah 54:16–17 shows us how God creates and governs the most powerful technologists. Reckoning with God's power over big tech is essential for many Christians who must resolve this obstacle before they can see and worship God for the tens of thousands of innovations they use every day.

Takeaways

It's hard to miss the emphatic point of Isaiah 54:16. It's reinforced by the declarative: "Behold." And then it's verified by the repeated personal pronoun and the particular verb for creation: "*I have created* [*bara*] the smith. . . . *I have also created* [*bara*] the ravager." Smiths and ravagers showcase God's unique work and his unmatched majesty.[34] Human tech is about him. The world's great innovators, who are working right now in a lab or factory or hangar or space station, exist by divine appointment. God is the genesis of human innovation and the creator of human inventors. Before there are makers, makers are made by God.

Soon we will look at how various industries emerged within human history. But first we should pause for a handful of takeaways from the tower builder, the blacksmith, and the sword wielder.

---

34 Young, *Book of Isaiah, Chapters 40–66*, 371.

1. God is not at the periphery of Silicon Valley; he's over it.

Isaiah's ravager thinks that he is powerful and untouchable because he does whatever he wants. But he doesn't know that the *whatever he wants* is fulfilling his personal purpose in the grand storyline of human history. God's sovereignty is most commonly demonstrated, not in contradicting human free will, but in working through human free will. He governs his creatures by governing their appetites, their *wants*. Sword makers *want* to be sword makers. Ravagers *want* to be ravagers. Thus, in the story of a particular smith and ravager (in Isa. 54:16), and more broadly in all the clamor of the nations (in Isa. 40:9–31), creatorship implies "total control over the actions of the creature."[35] In particular, "Yahweh has created and thus controls the one who makes weapons and uses them to cause devastation and carnage."[36] This control is wielded through the native impulses and desires of the heart.

On one hand, this is what it means to be saved. We must be given a new heart and a new soul that desire God.[37] We must want him, and in our sin we naturally don't. But our desire-driven nature also means that in the broader drama of human history, and particularly among the world's most powerful rulers, "the king's heart is a stream of water in the hand of the LORD; he turns it wherever he will" (Prov. 21:1). Kings act by heart, making decisions from that place where their powers of reason, feeling, and choice reside.[38] Thus, God directs each king and kingdom on earth by directing the king's native desires, his free will.[39] This is true for every king and

35 R. N. Whybray, *The Second Isaiah* (New York: T&T Clark, 1995), 57.
36 Paul R. House, *Isaiah: A Mentor Commentary* (Fearn, Ross-shire, UK: Mentor, 2018), 2:519.
37 Ps. 51:10; Ezek. 11:19; 36:26; Eph. 4:23.
38 John A. Kitchen, *Proverbs* (Fearn, Ross-shire, UK: Mentor, 2006), 463.
39 See also Gen. 20:6; Ex. 10:1–2; Ezra 1:1–2; 6:22; 7:27; Isa. 9:11; 13:17.

for every power player in Silicon Valley, as God "works all things according to the counsel of his will" (Eph. 1:11).

This understanding of God's work in the world was behind John Piper's 1993 address to a roomful of evangelical journalists. There he pleaded with them to become God-centered in their outlook on the world. God does not like to be taken for granted at any point of human history, Piper said—but journalists take God for granted every day in their reporting. They shouldn't. Why? Because of Isaiah 54:16, the only substantial mention of this text in Piper's entire, prolific ministry. Piper used the text to show journalists that behind every important piece of news, behind every important event, they will find God. To make his case, Piper cited the text: "Behold, I have created the smith who blows the fire of coals and produces a weapon for its purpose. I have also created the ravager to destroy." Then he said, "God is important because everything newsworthy—inventors, weapons, calamities—are created by God."[40] Think about the boldness of that statement. God is not distant from the headlines of Silicon Valley.

God has created the original creation; God will create the new heavens and the new earth; and God now creates the contemporary events we read about in the morning newspaper.[41] God is fully present and fully in control of the unfolding drama of man's technological possibilities.

But that's not the only way God governs innovation.

2. Any engineering feat of man can be squelched by God.

Journalists (and some Christians) seem content with a spiritualized world where God lives mostly in old stories and mythology, not

40 John Piper, "God Is a Very Important Person," sermon, desiringGod.org (May 11, 1993).
41 Young, *Book of Isaiah, Chapters 40–66*, 371–72.

inside the latest headlines coming out of tech culture. Humans with great knowledge and power get fooled into thinking God has been made irrelevant and powerless. We've outpowered him. But Scripture corrects this.

Because God creates the innovators, God can thwart the innovators. Forged iron cannot halt providence. God has no problem breaking iron restraints in the way of gospel ministry.[42] And on a much grander scale, he governs the stars and planets, the whole cosmic realm that gives us "an irresistible sense of machinery, clockwork, elegant precision working on a scale that, however lofty our aspirations, dwarfs and humbles us."[43] Yet within this elegant clockwork, God once paused the orbits to stop the sun's trek across the skyline.[44] This Old Testament event was not a long, dark eclipse of the sun or a literary myth about ancient Amorite omens. God accomplished a physical impossibility when he slammed on the orbital brakes without everyone on earth suddenly sliding around at 1,000 mph. How did he do it? I don't know. Cosmology has always been his plaything. But the result was that God interrupted the great wheels of a giant "machine" of the cosmos, halting the universe's orbit around the sun.[45] And if God can stop a cosmic machine like this, he can pause or stop any man-made machine.

Likewise, the early Industrial Age machines gave preachers a ready illustration for God's providence—as one big apparatus unified in a million moving parts. In a sermon on how providence

---

42 See Acts 12:10; 16:26.

43 Carl Sagan, *Pale Blue Dot: A Vision of the Human Future in Space* (New York: Ballantine, 1997), 98.

44 See Josh. 10:13; Hab. 3:11.

45 The word used by Jonathan Edwards, "'Images of Divine Things' 'Types,'" in *Typological Writings*, ed. Wallace E. Anderson, vol. 11, *Works of Jonathan Edwards* (New Haven, CT: Yale University Press, 1993), 61.

works, Spurgeon told his congregation to imagine entering the workshop of a machine engineer. All you see are piles of gears. They make no sense, scattered around on tables and carts. But let the engineer assemble the machine, and then you'll see each cog and gear and flywheel connected together. Right now, "you and I can never see more than parts of God's ways. We only see here a wheel and there a wheel; but we must wait till we get to heaven, then we shall see . . . that it was one piece of machinery, with one end, one aim, and one object."[46]

But this machine is not a robot. It's not autonomous. God did not wind up the cosmos like a watch and walk away. Quite the opposite. God is both the servo and the power source, says Edwards, "directing all the various wheels of providence by his skillful hand," and conspiring all things together, like "the manifold wheels of a most curious machine," all toward its final end: God's glory in his happy people, together with Christ in an eternal kingdom.[47] God powers all things, and turns all things according to his governance and design. The whole machine of providence works to a single, unified end according to God's plan, his glory, and his people's eternal joy.

But if God governs the machine of cosmic orbits, does he also govern the bioengineering lab, the epicenter of our greatest ethical dilemmas? In a survey of advances in biotech, theologian Hal Ostrander looked at human reproductive technologies, the promises of human gene editing, and even human cloning. He examined transgenic hybrids, the criss-crossing of genomes of

---

46 C. H. Spurgeon, *The Metropolitan Tabernacle Pulpit Sermons*, vol. 54 (London: Passmore & Alabaster, 1908), 498.

47 Jonathan Edwards, *A History of the Work of Redemption*, ed. John F. Wilson and John E. Smith, vol. 9, *Works of Jonathan Edwards* (New Haven, CT: Yale University Press, 1989), 525.

different animals to create super-pigs, super-cows, and super-salmon. Robotics, AI, and space travel were also fascinating to him. But for most of us, germline engineering, DNA editing, and human cloning mark the scariest and most nightmarish scenarios of technological futurism. But instead of stoking fear, Ostrander framed biotech within God's response to Babel. In Babel, man played God, not unlike some bioengineers today. And in the face of man playing God, God entered the story with punitive and protective measures, leading Ostrander to write: "God will allow us to go only as far as his providential measures in history and sovereign decrees from on high prescribe." In fact, "God will encapsulate, if not outright negate, immoral human efforts outside the range of his providential intentions." How so? Well it may mean that God has already set in place "scientific barriers that cannot be crossed, that is, physically instantiated limits with respect to misdirected genetic potentialities especially." God has set boundaries over many of his creational patterns, and perhaps he's already set scientific boundaries for what can be bioengineered in a lab. But whether his involvement in the world has already preempted the science, or whether he will repeat a Babel-like intervention, in all the futuristic predictions of bioengineering "there is mystery and comfort alike in the fact that God's sovereignty cradles our technological futures."[48]

Autonomous robots chauffeuring us around cities are likely in the near future. Autobots plotting a coup of mankind is more doubtful, but possible in the distant future. But autobots breaking free from God's sovereign governance will never appear. No maverick machine, bot, or genetic edit can thwart the governance

---

48 Hal N. Ostrander, "Technological Futures and God's Sovereignty: How Far Will We (Be Allowed to) Go?," *Southern Baptist Journal of Theology*, vol. 4/1 (2000): 52–54.

of God. He can allow them to run amuck and do real damage, but he can stop them too. God has set limits. He can, and does, and will continue to intrude upon our tech aspirations at his own will. What countless catastrophes has he stopped already?

3. God governs even over technologies of destruction.

To that prevailing idea that God's providence covers only butterflies, fluffy puppies, and healing technologies with no ill side effects, Isaiah 54 is a blast of reality. God reigns over every technology and every technologist. The point here is that God created sword makers and sword wielders for the express purpose of ravaging cities. He can wield destructive technologies through secondary causality, raising them up and deploying them for his own purposes.

So when Wendell Berry wrote that we love our neighbor by cultivating a craft, he made a very important point. We love our neighbors through the skills we learn and deploy in various fields. Where Berry and many others go wrong, however, is in assuming that God's use of human skills ends with the virtuous ones. The ravager proves that this assumption is untrue. Berry envisions a world where only "Christian vocations" are justifiable. "Is there, for instance, any such thing as a Christian strip mine?" he asks. "A Christian atomic bomb? A Christian nuclear power plant or radioactive waste dump? What might be the design of a Christian transportation or sewer system? Does not Christianity imply limitations on the scale of technology, architecture, and land holding? Is it Christian to profit or otherwise benefit from violence?"[49] No, it's not Christian to profit from violence. Christians will self-limit to virtuous vocations. But it's shortsighted to assume that God has

49 Wendell Berry, *Essays 1969–1990* (New York: Library of America, 2019), 525–26.

no use for a non-Christian ravager with a sword who will pillage for profit. Christians will self-limit our tech adoption, but there is no limit on God's intentions. He creates every technologist to serve his ultimate end for creation.

Berry is right that Christians will not deploy an atomic bomb in the name of Christ. Biblical convictions will always limit our tech adoption. But this is a long way from saying that God has no plan for an atomic bomb maker in his ultimate plan. If God created the smith and the ravager by divine intent, he creates the rocket scientist and the missile launcher today. Neither breaks free from God's power. Yes, he could stand back and allow a nuclear blast to humble man's techno-arrogance and to give a very real physical manifestation of the radiation burns of eternal wrath. But we can rest assured that God, who can save us from his own wrath, can save us from a nuclear holocaust. To be in the hands of God is a comfort for the godly and a terror to the godless.

Notice again I did not say that God *has to save us* from nuclear war to be good or sovereign. If he sees fit, God could allow us to unleash on ourselves our own judgment—devastating bombs, crippling cyberattacks, autonomous robot overlords, techno-barbarism, superbeasts, superviruses—any manner of new ravagers. By human hubris, God can raise up an innovator to unleash such a nightmare, and God would still be good and righteous and holy. Sometimes God ordains by his decretive or sovereign or secret will that which his moral character hates, but *only* if that particular evil serves a greater good that otherwise could not be obtained had the evil not occurred. God always has some infinitely good and wise purpose for what he ordains. However, we cannot always know what those greater goods and purposes are unless God reveals them to us. And most of the time he does not.

Again, God's governance over all these causes and effects will never make the ravager less guilty for his lust for blood or power or wealth. We know that God orchestrates technologies of human self-destruction without undermining his own holiness and without excusing the evil of the aggressors, because he did all this in the cross of Jesus Christ.[50] It was through metal technology that a blacksmith forged three long metal spikes and a sledge. And a ravager took these innovations and employed them to kill the author of life himself. In the words of the apostle Peter, in Acts 2:22–23:

> Men of Israel, hear these words: Jesus of Nazareth, a man attested to you by God with mighty works and wonders and signs that God did through him in your midst, as you yourselves know—this Jesus, delivered up according to the definite plan and foreknowledge of God, you crucified and killed by the hands of lawless men.

The rulers of Israel brought divine condemnation on their own heads for their sin. But it was also the "definite plan" of God. As Isaiah prophesied, Christ was "crushed" for our sins, and he was crushed by being pierced with metal nails.[51] An ancient blacksmith served God's ultimate design. "The LORD was *pleased* to crush Him" because it served his infinitely wise and benevolent purposes in the work of redemption (Isa. 53:10 NASB). Christ became sin on our behalf to take away judgment for our sins, and for it he was pierced by metal nails and a metal spear.[52] Christ was murdered by "lawless" ravagers, and his death was the "definite plan" of God—both

50 Acts 3:11–26.
51 Isa. 53:4–6.
52 2 Cor. 5:21.

statements are equally true.[53] God orchestrated our salvation in the death of his Son, through a sinful collusion between a nameless blacksmith and a Jewish and Roman system of injustice that played the role of ravager. God created this ravager for this purpose, to enact the greatest evil in human history, and all for the purpose of redeeming sinners like you and me.

The cross of Christ reminds us of two important points. First, human hatred of God dominates every age. "The men who built the city against God [in Babel] . . . had the same hatred as that which possessed the men who nailed the Lord Jesus Christ to the cross."[54] This same anti-God rebellion is alive today. Second, and despite this fact, God continues to use human technologies both to judge and to bless humanity. Babel and Golgotha force us to see the complexity of God's sovereign relationship to human innovation.

Every inventor, every invention, every use of every invention, and every outcome from every invention—they all fall under the Creator's disposal. This includes the nameless blacksmiths whom God created to fire the coals and fashion a hammer, three iron spikes, and a spear, so nameless ravagers could pierce into the flesh of the Savior.[55] Mankind sought to dethrone God through construction technology in Babel. And mankind sought to dethrone God through metal technology at Calvary. But God hacked human tech, taking what was intended for evil and turning it for good—the redemption of his church.

So when the ravager ravages, will humanity see an "accident" or a misuse of technology? Or in the fallout of our technologies, will we

---

53 Acts 2:22–23.

54 Donald Gray Barnhouse, as quoted in Brad Waller, "For the Church: Discipling Every Age," *Tabletalk*, February 2014 (Sanford, FL: Ligonier Ministries, 2014), 66.

55 John 19:34; 20:25.

hear the chastening voice of the Creator, bringing down judgment on the hubris of man? In other words, don't fear the genetically modified superspecies. And don't fear the maker of a genetically modified superspecies. Fear the one who can raise up a technologist to manifest the destructive powers of innovation. No destruction will befall our world except those that God governs for his ultimate purposes.[56] In other words: fear God, not the technicians.[57]

4. Energy for human innovation comes from the Spirit.

Unlike the animals around us who seem content to live on the earth, man uses technology to try to escape this fallen planet, to search for a better world. Babel reveals our fundamental uneasiness in this place and time, because, as one philosopher put it: "Man's being and nature's being do not fully coincide. Because man's being is made of such strange stuff as to be partly akin to nature and partly not, at once natural and extra-natural, a kind of ontological centaur, half immersed in nature, half transcending it."[58] Animals feel little need to innovate to endure this planet. But we keep inventing and evading and trying to escape.

Much of technology emerges from this desire to transcend nature, to be freed from fallen circumstances around us. Technology is a way for us to say: "We don't really fit here. We must escape. We should storm heaven or at least attempt to colonize Mars."

Babel was man's first attempt to run away from this planet, to launch himself into the blue abyss of the atmosphere, and to create his own entrance into heaven itself. All of our space exploration echoes something of Babel (as we will see later). But

56 Amos 3:6.
57 Matt. 10:28.
58 José Ortega y Gasset, *Toward a Philosophy of History* (New York: Norton, 1941), 111.

more fundamentally, the use of technology and engineering to transcend this earth is a manifestation of human self-definition and personal identity shaping. These efforts make the engineering feat of Babel a spiritual ambition. Whether we are aware or not, we seek identity in our technologies. Engineering taps into a human spiritual impulse.

Before we innovate, we imagine. Our technological imagination comes from a "spiritual nature" within all humans, not just Christians.[59] Calvin was unafraid to call it *spiritual*, a type of fruit of the Spirit's presence in the lives of even non-Christians who are not indwelt by the Spirit. God "fills, moves, and quickens all things by the power of the same Spirit [that saves], and does so according to the character that he bestowed upon each kind by the law of creation." The Spirit that saves is the same Spirit that causes the technological gifts of man to flourish. Calvin writes, "If the Lord has willed that we be helped in physics, dialectic, mathematics, and other like disciplines, by the work and ministry of the ungodly, let us use this assistance. For if we neglect God's gift freely offered in these arts, we ought to suffer just punishment for our sloths."[60] All innovation is capital-*S*, Spiritual, the work of the Spirit himself.

A technological genius seeking transcendence gives evidence of the Spirit's work of common grace. But that same person may be spiritually dead. A vibrant technological mind says nothing about the vibrancy of the soul. And yet, as Calvin makes clear, the same Spirit is at work—in the latest smartphone release or in the con-

59 Stephen Charnock, *The Complete Works of Stephen Charnock* (Edinburgh: James Nichol, 1864–1866), 1:265.

60 John Calvin, *Institutes of the Christian Religion*, ed. John T. McNeill, trans. Ford Lewis Battles, Library of Christian Classics (Louisville, KY: Westminster John Knox, 2011), 2.2.16.

version of a soul. God inspires and motivates innovators through his Spirit. From those innovations the church finds beneficial gifts to adopt.

5. Each innovator exists by divine appointment.

Many of the sharpest Christians, who rightly celebrate God's providential governance over all things, tend to wrongly assume (in practice) that his reign ends somewhere around the boundary lines of Silicon Valley. In reality, innovators—both virtuous ones and nefarious ones—are created by God. Scripture protects us from the myth that God is trying his best to stifle and subdue the unwieldiness of human technology. No, for his own purposes God *creates* blacksmiths and warriors, both welders and wielders of new tools. Our most powerful innovators exist by divine appointment.

More troubling, many of the world's most powerful technologists imagine that they have transcended their need for God. And it is their common agnosticism or atheism that explains why Christians today often adopt a negative view of technology. The godlessness of Elon Musk reminds us that the closer you approach Silicon Valley, the fewer Christians you'll find. The percentage of professing evangelical adults in the US (25.4 percent) drops in California (20 percent) and plummets in San Francisco (10 percent). And the percentage of adults who read Scripture at least once a week in the US (35 percent) sinks in California (30 percent) and plunges in San Francisco (18 percent).[61] We assume that God must be withdrawn from such a place. But Isaiah corrects this assumption. The pagan societies where the ancient smith and ravager operated make San Francisco look like it's part of the Bible Belt.

---

61 Pew Research Center, "Religious Landscape Study," pewforum.org (2014).

The rejection of God and the accumulation of innovative brilliance doesn't give you the power to operate apart from God, like a queen on a chess board who thinks she can move anywhere she wants, impervious to the Master's ultimate plan. Your innovative brilliance is *how* God is choosing to wield you in the world. If you find in yourself a zealous impulse to forgo sleep in order to make new innovations, that zeal was implanted inside of you, by the Spirit, for a greater ultimate purpose that far exceeds what you can see.

God created the power-class innovators of the ancient world who trafficked in the world's most dangerous and destructive tech. Only he does this. Again, the Hebrew word for *create* (*bara*) "is used in the Old Testament only of divine action, to express those acts which by their greatness or newness (or both) require a divine agent."[62] In Isaiah 54:16 this word is repeated in front of the smith and the ravager. Makers and wielders of war tech, in any generation, require a maker. God is their maker. His free and unrestrained activity among us, and his ongoing creative acts within this world, are on display even today, as he creates new innovative makers and raises up new wielders of potent technologies. God populates Dubai and Bengaluru and Silicon Valley and Silicon Alley and Silicon Prairie with their most powerful innovators.[63]

---

62 J. A. Motyer, *The Prophecy of Isaiah: An Introduction and Commentary* (Downers Grove, IL: InterVarsity Press, 1996), 66. See also 378.

63 If the dramatic origin of smiths and ravagers sounds uber-spiritual, it actually illustrates a larger dynamic in the wisdom of God. As one theologian put it: "God does not know things because he came to know them through discovery and deduction. God knows all things because he knows himself, and all things are from him, through him, and to him" (Samuel D. Renihan, *Deity and Decree* [self-published, 2020], 70). This is true of all creation and every creature. It includes all the gifts and purposes of your life and mine. The depth of God's wisdom in knowing everything about you and me is not about his ability to search and study the independent lives we lead. No. God's wisdom of his creatures has everything to do with his sovereign providence, which creates and positions and wields each of his unique creatures according to his own design. Therefore, the "depth of the riches and wisdom and

God's claim as the creator of the world's most powerful industry is a cosmic checkmate for brilliant inventors today who imagine that their powers of innovation have made the Creator irrelevant. No! God wields you as you innovate. Your innovations serve his end. Elon Musk claims to have put in 120-hour work weeks on occasion but says that he's normally down to a "pretty manageable" 80- to 90-hour work week.[64] Why so many hours? Because God created him to work like a farm mule. For all I know, Musk's motive is for wealth or power or prestige. And it doesn't matter. God created Elon Musk to be Elon Musk. Whether you love God, hate God, or ignore God; and whether you seek to meet the needs of humanity in your work, or whether the only thing that gets you out of bed each morning is the promise that you're going to plunder this world of as much wealth as you can, with a sword or a startup, God wields you for his final purposes. God made you for an end that he set in place. And if technology and innovation are your field, this is where you fulfill that end. God creates makers, and he creates wielders of technology for beautiful and healing purposes. And he creates makers and wielders of technology for gross and ravenous purposes. He's the potter, as we are told in Romans 9. He may use you to discover the genetic cure for cancer, or he may use you to weaponize a superravager, but he disposes of every innovator as he pleases in his wisdom. We are each accountable for our volitional

---

knowledge of God" over his creation is because all of creation is "from him and through him and to him" (Rom. 11:33–36). All conscious creatures "were created through him and for him" (Col. 1:16). God is "over all and through all and in all" (Eph. 4:6). The Creator, even today, "works all things according to the counsel of his will" (Eph. 1:11). God governs each of his creatures toward good ends (directly). And he governs all sin and evil (indirectly). But he really governs all things, including the vocational decisions of his creatures.

64 Eric Johnson, "Full Q&A: Tesla and SpaceX CEO Elon Musk on Recode Decode," vox. com (Nov. 2, 2018).

decisions and our sins. But make no mistake—each and every one of us finally fulfills the Creator's purpose for our lives.

Isaiah 54:16 destroys every shred of assumption that my powers of innovation make God more distant and less relevant to my life. Only a fool would come to that conclusion. It's exactly the opposite. God makes innovators. They exist by his design alone. By them he governs humanity's present, and its future.

6. God controls the future by creating the innovators of that future.

Wendell Berry and others assume that Christianity is a useful ethical worldview for technological asceticism. Our faith is certainly a framework to help us make life decisions. But it's much more than that. Christianity is the revelation of the sovereign God of the universe, who made Thomas Edison and Steve Jobs and Elon Musk for his own purposes—beyond whether or not their innovations should be adopted into the lives of Christians.

Even more starkly, one of the fundamental tenants of open theism says that God cannot know all things timelessly but must learn and discover as things take place.[65] So when Christians try to bring God into the world of modern innovation, many slip into the faulty logic of open theism. Every once in a while, they think, God turns his attention in our direction to see what we are inventing now, to learn what innovations are possible, to see where our techno-trajectory is headed next. God bends down to study Silicon Valley in order to self-discover and self-improve himself as time goes on because his future self-development hinges on future human innovation.

This caricature is backwards nonsense. God makes innovators. Technological advance is evoked by the Spirit under the sovereign

---

65 John M. Frame, *No Other God: A Response to Open Theism* (Phillipsburg, NJ: P&R, 2001), 23.

sanction of God. Which means that not only is God sovereign over the sword maker and the sword wielder; God has control of the future. This is Isaiah's point. Isaiah wants us to comprehend that the living God of the universe controls the future, because he is the one who creates the inventors of that future. No human invention teaches God. And no innovator catches God by surprise. He makes each innovator. That's profound, providential logic. Or to speak more specifically: in his wisdom and for his purposes, God intended for the world to have iPhones; therefore he created Steven Paul Jobs to be born on February 24, 1955. God governs the future by creating the blacksmiths and the innovators who shape that future. Much needs to be said about how industries are born, and we will get to that discussion later.

7. God reigns over technologies that heal.

Isaiah 54 makes a clear point by focusing on new technologies that are made to plunder cities. But think of the logic. If God reigns so powerfully over technologies that ravage, how much more does he reign over virtuous technologies that heal? Think of laser eye surgeries, insulin pumps, bilirubin lights, nebulizers, ventilators, kidney machines, pacemakers, and heart defibrillators. God gets full credit for innovations that heal. And there are thousands of reasons to praise the Spirit for these gifts and for the vast array of innovations he has given us to use every day in our tech-saturated lives.

8. God focuses his attention on the actors of technology.

Tech ethics is not binary. We can't drop a gadget into an ethics machine and wait for it to come out the other side, stamped as "virtuous" or "sinful." If we determine that the smartphone is inherently sinful, then to touch it would be a sin. But Scripture won't allow such a clean dichotomy.[66] In reality, the inherent sinfulness or virtue of a given technology is often vague until an actor wields the tool with intent.

Melvin Kranzberg, a historian of technology at Georgia Tech, once penned six laws of technology. Law number one is worth printing on a bumper sticker for your electric car: "Technology is neither good nor bad; nor is it neutral."[67] This is a helpful way to speak of innovation. To speak of technology as "morally neutral" really gets us nowhere. We can have meaningful discussions, however, about the tech we use and the motives that drive us. Additionally, I find it difficult to imagine any technology that cannot be used for both good *and* evil—to tar an ark *and* to tar a tower. In either case, neutrality isn't a governing category that will help us progress ethically, and that is because innovations continue to come from the hands of rebel sinners who reject God.

No technology is ambivalent; each one comes with certain biases and tendencies. The true challenge of ethics is not in determining which technologies should be made possible but in determining how those new possibilities are wielded. Thus, Scripture puts the emphasis not on the technology, but on how those innovations are used.

9. God's governance over all human innovation is meant to comfort.

In God's plan, technology will serve the ultimate flourishing of God's people. That's the takeaway of Isaiah 54. God controls the

---

66 Col. 2:20–23.
67 Melvin Kranzberg, "Technology and History: 'Kranzberg's Laws,'" *Bulletin of Science, Technology, and Society* 15 (1995): 5.

future by creating inventors who shape the future. Why? To what end? Look at the second half of verse 17. God creates both the weapon inventors and the weapon wielders, and he uses them and limits them, all for the sake of "the heritage of the servants of the LORD" and for their ultimate "vindication." So who are these vindicated servants?

Two different servants factor prominently in Isaiah. One is God's people, his remnant. But there's a second servant, a solo character. We must know both of them. Rewinding a few verses into chapter 53, we arrive at an incredible chapter, the text where we are told that the Lord has a servant, and that this single servant, whoever he is, will be crushed for our iniquities.[68] Look near the end of this remarkable passage, at Isaiah 53:11: "Out of the anguish of his soul he shall see and be satisfied; by his knowledge shall the righteous one, my servant, make many to be accounted righteous, and he shall bear their iniquities." The "my servant" (singular) is a character who will be crushed for sin. So that the "many servants" (plural) will be justified.[69] God will justify his many servants, through the blood of one servant—his Son Jesus Christ. A Son will be murdered by a blacksmith and a ravager. And it's all part of God's plan to vindicate his people and bless them with eternal joy.

Pieced Together

So let's put all the pieces together. God creates inventors and populates Silicon Valley with powerful tech gurus. And God will ensure that certain technologies are held in check through cross-cultural tensions that he has coded into humanity (beginning at Babel) and by exerting direct sovereign power to interrupt and squelch

68 Isa. 52:13–53:12.
69 House, *Isaiah: A Mentor Commentary*, 2:520–21.

or hack human innovation when needed (proven in Babel and Calvary). Therefore, the reason why no weapon that is fashioned against God's people will stand is that God is sovereign over each and every weapon. He makes its maker, and he makes its wielder. Every weapon, every human innovation—even the most destructive human innovations—serve God's redemptive plan. Every motivated scientist and every human innovation and every wielder of those innovations all operate in subservience to God's love for his people. Why is this important? Because God's people need this assurance in a fallen world. God is unfolding his plan, which is the ultimate good and the final vindication of a blood-bought people, paid for in full by the death and resurrection of Jesus Christ!

Christ is building his church, and the iron gates of hell—all the mythic blacksmith technologies combined, using all the available iron in the universe—will never prevail against his bride.[70] Every power inherent in the technologies of man submits to Christ's plan for his bride. The locking chains on Paul will hold fast if God permits it.[71] But the locked prison door will swing open if God commands it.[72] Bound or free, no human restraints will stop the church apart from God's governing permission. Every metal of the earth, hardened by fire, will obey the will of the smith's maker. As the apostle Paul said, more than seven hundred years after Isaiah: "If God is for us, who can be against us?" Nothing. Not the iron. Not the tech maker. Not the tech wielder. We know that God is for us, because God sent his Son! He will stop at no lengths to redeem us. For "neither death nor life, nor angels nor rulers, nor things present nor things to come, nor powers, nor height nor depth, nor

70 Matt. 16:18.
71 Acts 28:20; Eph. 6:20; Phil. 1:13.
72 Acts 5:17–26.

anything else in all creation"—not even the collected technological powers of humanity—"will be able to separate us from the love of God in Christ Jesus our Lord" (Rom. 8:31–39).

This is God's relationship to technology.

So What's the Tar For?

God calls each of his children to make faith-based decisions on how we use the available technologies in this world. Tar is the metaphor. The question each of us must answer is this: What's the tar for? Do we waterproof as an act of faith in God or as an act of rebellion against him? Noah used tar because he was on a journey to find a future promise. Tar sealed his boat and tightened his faith in God. But the Babelites misapplied the tar, because in unbelief they waterproofed their fired-hardened bricks to construct a permanent utopia of self-glory.

So, returning to where we started, the first dozen chapters of Genesis offer us two major technology stories. First is a story of God's redeeming grace in and through human technological innovation (the ark, the first ever massive ship). Second is a second story of humankind manifesting its sin through technology (the city-tower of Babel). The first use of technology was based purely on faith. The second use of technology was based entirely on unbelief. Technology used by a heart of flesh versus technology used by a heart of stone. And God governed both.

Now to this point, our theology of technology is off to a good start. But we have not addressed questions about how industries are born, or where our tech comes from materially, or how our tech relates to the possibilities within the created order. Eventually we will need to understand what innovations can never do for us. And we must determine what technologies are destructive or redemptive, which ones we should adopt or reject. Because at some point we must ask: What's the tar for? You can use tar to waterproof your ark (in faith). Or you can use tar to waterproof your tower (in unbelief). But our discussion of ethics must wait.

3

Where Do Our Technologies Come From?

EX NIHILO IS A PERFUME COMPANY based in Paris. On their website, in a written manifesto, they rightly define the Latin phrase *ex nihilo* as "creating out of nothing." But just two sentences later, they announce a corporate commitment to make elegant fragrances for their customers from "the most exclusive perfumery materials."[1] (Ha!) There's the rub with human innovation. We talk about creating "from scratch," but we can't. All we can do is mix and match, add and subtract. We are confined by what's available.

So when Steve Jobs "invented" the iPhone, he really didn't invent much of anything. He simply recondensed dozens of known technologies—a digital music player, a cell phone, a calendar, a Rolodex, and a web browser—all into a digital PDA with a touch screen. Human innovators are really just discoverers, separators, mergers, replicators, copyists, and refiners. More generally, technology is like playing in a sandbox that someone else made. We create

1   "Manifesto," ex-nihilo-paris.com (n.d.).

nothing out of nothing—not cars, not smartphones, not perfumes. We discover and copy and paste and rearrange. We operate within boundaries set by someone else.

Our Bibles tell us who built this sandbox, in Genesis 1:1: "In the beginning, God created the heavens and the earth." God made everything, and he made everything from nothing. The earth he made from nothing. And everything else, in the sky and in space, in the animals and in man, he made from nothing. *Bara.*

Unapproachable Light

The apostle Paul tells us that this Creator "dwells in unapproachable light" (1 Tim. 6:16). God is undivided. He is "pure, enduring, self-original light."[2] God is entirely self-sufficient and self-satisfied in his eternal being in the absence of all that is not God. He casts no shadow.[3] He has no lack. He needs nothing outside of himself. Man cannot create God, overthrow God, improve God, or limit God. God's existence is determined by nothing in creation. He is not waiting to be comprehended by us in order to know himself. Every sermon spoken or theology book published adds nothing to his self-understanding. Human innovation has nothing to add to his fullness.

In this glimpse of unapproachable light, when *all that is not God* gets thrust aside, and all that remains is God—when no angels or humans or animals or galaxies are in the frame—we get a snapshot of God, self-existing in his lavish fullness, a glorious self-sufficiency that transcends and predates time, space, and everything he has made.[4] Nothing outside of God adds to him. Nothing outside of

---

2 John Webster, *The Domain of the Word: Scripture and Theological Reason* (New York: T&T Clark, 2012), 57.
3 James 1:17.
4 John 17:5.

God completes him. His unapproachable radiance pushes back everyone else and everything else so we can marvel at his self-sufficiency. He needs no created thing. The making of creation does not lessen God.[5] "The existence of creation adds nothing to God, and in its absence God would be undiminished," writes theologian John Webster. "God is in himself infinitely happy, in need of nothing from the creature."[6] In the words of Jonathan Edwards, "God is infinitely happy in the enjoyment of himself"—God as Father, delighting in his exact image, his Son, through the mutual love of the Holy Spirit. The triune God is white-hot self-delight, something akin to nuclear fusion in a reactor, producing "a most pure and perfect energy in the Godhead, which is the divine love, complacence, and joy."[7]

And yet this same God chooses to not self-limit his self-sufficient glory as *only* unapproachable. The material universe is "an explosion of God's glory," a purposeful detonation outward as "perfect goodness, beauty, and love radiate from God and draw creatures to ever increasingly share in the Godhead's joy and delight."[8] God's infinite happiness within himself erupts and creates a universe entirely outside of himself—and he invites this universe back to participate in his very life. That the world exists at all (that you and I exist at all), can be explained in no other terms than the sheer intentional generosity of God. God has no needs, so creation cannot complete him. The

5   John Webster, *God without Measure: Working Papers in Christian Theology, Vol. 1: God and the Works of God* (New York: T&T Clark), 160.

6   Webster, *God without Measure*, 115–26.

7   Jonathan Edwards, *Writings on the Trinity, Grace, and Faith*, ed. Sang Hyun Lee and Harry S. Stout, vol. 21, *Works of Jonathan Edwards* (New Haven, CT: Yale University Press, 2003), 113.

8   George M. Marsden, *Jonathan Edwards: A Life* (New Haven, CT: Yale University Press, 2004), 463.

material universe is a pure, generous gift of surprise, a succession of "Let there be's!"—divine commands granting permission for creation to exist, like some long-ago-made secret that can no longer be held back, a world coming into existence as an utterly astonishing reality owing only to the overflowing happiness of God.[9] You, me, angelic beings, this planet, this universe, the tech under your thumbs and on your wrist—none of it is *necessary* to God's radiant life and happiness. And yet here we are, tech-advancing creatures created to be drawn into the fellowship of the self-sufficient God.

Matter cannot explain the origin of matter. So we gaze at the material world and ask: Why is there something rather than nothing? And that answer is found in God. *He is.* He is gushingly delighted in himself. Therefore, all things exist in total originality. God requires no raw materials to create, no preexisting patterns to follow, no base elements to form compounds, no originals to copy. This is because he "himself is his own pattern and copy in his works."[10] Originating from within this white-hot Trinitarian delight, his unapproachable light, every element in creation is patterned inside God and generously commanded into existence.

Oxygen, silicon, aluminum, iron, calcium, titanium, hydrogen—students memorize their properties in chemistry class because these elements build other compounds, but God created each atom from nothing. *Bara.* The material world exists by God's intent, down to quarks, electrons, muons, and photons, the "basic constituents of matter" that mark off "the very boundary of being in its creation out of nothing."[11] By his wisdom, God founded everything

9   See the fifteen instances of "let" in Gen. 1:3–26.

10   Stephen Charnock, *The Complete Works of Stephen Charnock* (Edinburgh: James Nichol, 1864–1866), 2:107.

11   Thomas F. Torrance, *The Christian Frame of Mind: Reason, Order, and Openness in Theology and Natural Science* (Eugene, OR: Wipf & Stock, 2015), 55–56.

under our feet and everything above our grasp, from the bottom of the ocean to the top of a thundercloud.[12] Every physical constant we discover inside creation points to the glory of God, who interrogated Job out of the whirlwind with a list of marvels that the most acclaimed nature documentaries try to capture—storehouses of snow and tracks of thunderbolts, gates of deep darkness and songs of morning stars, the leaping of locusts and the oddness of ostriches.[13] God alone set every boundary in place before humans existed. We had zero influence on how this earth and the universe were patterned.

We came into being as stewards of a preset creation. So our chemists can separate, replicate, refine, and merge what already exists. "But when God speaks a powerful word, *nothing* begins to be something."[14] *Nothing* is not God's raw material. God doesn't need *nothing* in order to make something. *Nothing* is not a big bang or a black hole. *Nothing* is nothing. Pure absence. Total nonexistence. Stephen Charnock writes, "A greater distance cannot be imagined than the distance between *nothing* and *something*, that which *has no being* and that which *has being*. And a greater power cannot be imagined than that which brings something out of nothing."[15] For *nothing* to give way to *something* is beyond the grasp of our human imagination, and certainly beyond our innovative skill.

Creatio ex nihilo means that the material universe "does not derive from any necessity in God and does not have any intrinsic necessity in itself." It exists by "pure freedom." Within itself, creation "contains no reason in itself why it should be what it is and why

12 Prov. 3:19–20.
13 Job 38–39.
14 Charnock, *Complete Works of Stephen Charnock*, 2:111.
15 Charnock, *Complete Works of Stephen Charnock*, 2:128; emphasis added and language modernized.

it should continue to exist."[16] Nonbeing came into being, a new being distinct from God's own being.[17] This is why we exist. This is why creation exists. This is why technological possibility exists. All our science and innovation are the result of the nonnecessary, pure freedom, intentional design, and sheer generosity of God.

Hold that cosmic thought as we hop back down into human history.

Adam and Eve

In the biblical story, the first humans were created as stewards. They were sinless and yet also incomplete. The Creator intended for Adam and Eve to reach a future glorification, a future not fully given to them in their unfallen state. Sinless humanity was God's crowning work, but, like creation, it was a work in progress. A future glorification of these beings had to come later. And an identical paradigm operates within the globe. This world was created sinless and endowed with potential. Creation is not necessary; it's the product of God's freedom and his immense love, and yet "it is a love that leaves the creature something to be and do," to cultivate the creation.[18] God created humans to discover the patterns of creation and to develop new technologies as a result. And this is what they did.

So where do new industries of human innovation come from? To answer this important question we rewind back before Babel, before the flood and Noah, back to the first couple, and after their fall into sin. Genesis 4:1–2.

16 Thomas F. Torrance, *Divine and Contingent Order* (Edinburgh: T&T Clark, 1998), *vii, xi.*
17 Herman Bavinck, John Bolt, and John Vriend, *Reformed Dogmatics: God and Creation*, vol. 2 (Grand Rapids, MI: Baker Academic, 2004), 416.
18 Colin E. Gunton, *Christ and Creation* (Milton Keynes, UK: Paternoster, 1992), 77.

¹ Now Adam knew Eve his wife, and she conceived and bore Cain, saying, "I have gotten a man with the help of the LORD." ² And again, she bore his brother Abel. Now Abel was a keeper of sheep, and Cain a worker of the ground.

God made the first man, Adam. Eve claims to have made the second man, Cain, with some help from God.[19] God and woman are joined together in the joyful fellowship of cocreation.[20]

Cain's name means "produced," a hint at his origin and a foreshadowing of his lineage to come. Abel's name means "breath," a foreshadowing of his life's brevity. They are not simply new brothers; they are two new lineages in human history.

So Adam and Eve bore two sons. Cain is the oldest. And if you know your Bible well, older brothers don't often relate well to their siblings. Additionally, there may be an occupational rivalry too: shepherds versus farmers. In a postagrarian society like ours, we don't get the full sense of the conflict. Abel wielded dominion over animals and raised sheep. Cain wielded dominion over the soil and raised grain. They were both vocationally proficient. But their story, older brother versus younger brother, is tense from the start.

³ In the course of time Cain brought to the LORD an offering of the fruit of the ground, ⁴ and Abel also brought of the firstborn of his flock and of their fat portions. And the LORD had regard for Abel and his offering, ⁵ but for Cain and his offering he had no regard. So Cain was very angry, and his face fell.

19 For Adam's response, see the "Obama come on what GIF."
20 Umberto Cassuto, *A Commentary on the Book of Genesis: Part I, From Adam to Noah* (Genesis I–VI 8), trans. Israel Abrahams (Jerusalem: Magnes Press, 1998), 201–2.

Cain became the first human to ever bring an offering to God. He owns the patent for the first human religious sacrifice, the enigmatic originator of the whole religious offering system.[21] Abel followed. Cain brought some crops; Abel brought the first of his flocks. Abel's offering was accepted. Cain's was scorned.

In this passage, we are not told why God accepted Abel's offering (of animal fat) and rejected Cain's offering (of grain). Both offerings were legitimate. For that answer we turn to the New Testament. There we are told that God accepted Abel's sacrifice because Abel trusted God. Cain didn't.[22] Abel did not put his self-confidence in his ability to breed and raise animals and accumulate wealth. At the core of his life, he lived for a future reward. Abel's faith sanctified his rudimentary breeding insights. He looked to God to deliver what his career could never give him. This is what it means to glorify God in our creating. It's about our heart, our allegiance, and where we look for our ultimate hope and final security.

Cain pledged no such allegiance; he did not consecrate his life and innovation to God. Perhaps he lived for this present life, trusting in his originality and prosperity. On the outside, Cain looked a lot like his brother. He looked like a philanthropic farmer. But God didn't have Cain's heart. That's the difference. Innovation directed to God's glory versus innovation pursued for worldly success—the ambitions may look parallel, but their paths diverge eternally. The elder brother grew jealous.

⁶ The LORD said to Cain, "Why are you angry, and why has your face fallen? ⁷ If you do well, will you not be accepted? And

---

21   Rabbi David Fohrman, *The Beast That Crouches at the Door* (Baltimore, MD: HFBS Press, 2012), 97–98.

22   Heb. 11:4.

if you do not do well, sin is crouching at the door. Its desire is contrary to you, but you must rule over it."

Cain was crushed. But God met him, encouraged him not to give up but to press on in his devotion, and urged Cain to resist giving in to his angry jealousy. The plea fails.

> [8] Cain spoke to Abel his brother. And when they were in the field, Cain rose up against his brother Abel and killed him.

In the first-ever premeditated murder, Cain waited to be alone with Abel in a distant field. Perhaps Cain used a farming tool as a weapon. We don't know. The forensics are long swallowed in dust, but the offense is recorded in Scripture. Cain killed his brother Abel and left his body behind for his shed blood to be swallowed by the soil.

Cain and Abel were hostile brothers, "engaged in a fratricidal struggle that ends in the death of the *best* one of them," a nightmarish scenario that is the tragic story of mankind.[23] The evil inside the drama of human history is haunting, and the resulting sorrow of this scene is thick, captured by many classic paintings of the horrific moment when Adam and Eve discovered the body of their dead son—the chilling first spectacle of human death brought into the world by his parents' first sin.

In the agonizing drama, God stepped in and spoke to Cain.

> [9] Then the LORD said to Cain, "Where is Abel your brother?" He said, "I do not know; am I my brother's keeper?" [10] And the LORD said, "What have you done? The voice of your brother's

---

23 Jordan B. Peterson, "Biblical Series V: Cain and Abel: The Hostile Brothers," youtube.com (June 27, 2017).

blood is crying to me from the ground. [11] And now you are cursed from the ground, which has opened its mouth to receive your brother's blood from your hand. [12] When you work the ground, it shall no longer yield to you its strength. You shall be a fugitive and a wanderer on the earth."

A murdered body in a field calls for blood.[24] For first-degree murder, Cain deserved execution on the spot. But his life was spared. He would live but he would be offered no forgiveness, no salvation, no grace. His farming life was over. The ground, already cursed, now refused to flourish for Cain. It would give him no wealth and no more offerings. Cain is expelled as a fugitive in perpetual exile to carry his guilt and agonizing curse all his remaining days on earth.

But God is not done with Cain.

[13] Cain said to the LORD, "My punishment is greater than I can bear. [14] Behold, you have driven me today away from the ground, and from your face I shall be hidden. I shall be a fugitive and a wanderer on the earth, and whoever finds me will kill me." [15] Then the LORD said to him, "Not so! If anyone kills Cain, vengeance shall be taken on him sevenfold." And the LORD put a mark on Cain, lest any who found him should attack him. [16] Then Cain went away from the presence of the LORD and settled in the land of Nod, east of Eden.

Not only is Cain *not* executed, but God shields his life and marks Cain in some obvious way that says: Don't mess with this guy. Don't assault him. Don't harm him. Don't kill him.

24 Deut. 21:1–9.

So why is Cain's lineage so carefully protected, when this murderous patriarch is so worthy of death? Your answer to this question will eventually determine whether you become a tech pessimist or a tech optimist. You'll see why if we keep reading.

> [17] Cain knew his wife, and she conceived and bore Enoch. When he built a city, he called the name of the city after the name of his son, Enoch.

In the time after Adam and Eve and before Noah's ark, Cain built the very first city mentioned in Scripture (a preflood predecessor to postflood Babel). Cain, exiled from his fields, turned to urban planning and designed what became the very first named city on earth. His father and mother, Adam and Eve, named the animals and their own children. Cain named his city. He made something inanimate and named it and claimed it as his own innovation. His innovation inaugurates both city building and intellectual property.

Now, as a reminder, we are asking: Where does human technology come from? And we return to Cain's lineage for some clues.

> [18] To Enoch was born Irad, and Irad fathered Mehujael, and Mehujael fathered Methushael, and Methushael fathered Lamech.
> [19] And Lamech took two wives. The name of the one was Adah, and the name of the other Zillah.

The genealogy picks up speed to arrive at Cain's great-great-great-grandson Lamech, inventor of polygamy. He and his two wives, Adah and Zillah, bear four notable children.

[20] Adah bore Jabal; he was the father of those who dwell in tents and have livestock.

Abel kept sheep, but Jabal raised various livestock. This is novel. Jabal will also invent mobile housing, habitable textiles, and animal breeding, basically what we now call the rudimentary beginnings of genetics. And Jabal had a brother.

[21] His brother's name was Jubal; he was the father of all those who play the lyre and pipe.

Jubal was a genius who invents music and strings and wind instruments simultaneously. The music industry is born to lead society in song. Then along comes a third brother.

[22] Zillah also bore Tubal-cain; he was the forger of all instruments of bronze and iron.

The industry of tool making begins with Tubal-cain, a man skilled to fashion and sharpen swords and new farming tools. Both the Bronze Age and the Iron Age are born simultaneously.
Then the three brothers get a sister.

The sister of Tubal-cain was Naamah.

The sister's mention in this ancient genealogy is surprising. She must have been noteworthy. The only evidence we have is her name, which suggests the sweetness of her feminine voice. Maybe Naamah was the first professional female vocalist.[25] That would fit with Jubal's invention of the music industry.

25 Nahum M. Sarna, *Genesis, The JPS Torah Commentary* (Philadelphia: Jewish Publication Society, 1989), 38.

But following this genealogy of remarkable innovators, father Lamech broke out with the first recorded song in human history: an angry, arrogant, self-centered, gangsta-rap track about self-preservation and vengeance (4:23–24). Lamech had apparently killed someone, and he would kill many more in vengeance. Cain's murderous spirit lived on; indeed it seemed to amplify. Rapid tech advances never advance ethics at the same rate.[26] For now, Cain's lineage continued, but not as a story of spiritual inheritance. His lineage is the cursed and spiritually dysfunctional story of innovative brilliance. "Cain's family is a microcosm: its pattern of technical prowess and moral failure is that of humanity."[27]

Then—record scratch—the story rewinds to Adam and Eve, who bore another child, Seth, Abel's replacement, and the forefather of an engineer to come, a man of faith named Noah.[28]

Cain's Lineage

We will return to Seth and Noah's lineage later. For now, we must appreciate again that God could have executed Cain for the premeditated, cold-blooded murder of his brother. He didn't. Instead, God protected Cain for a specific purpose. The genealogy makes clear that God was not simply protecting a man; God was safeguarding a lineage.

In Cain's spiritual descent, the inaugural city emerged. From that city, through Lamech and his two wives, human innovation spiked. All of a sudden, in rapid succession, the forefather of durable textiles and mobile housing and animal breeding was on the scene. And

---

26 Cassuto, *Commentary on the Book of Genesis*, 244.

27 Derek Kidner, *Genesis: An Introduction and Commentary*, Tyndale Old Testament Commentaries (Downers Grove, IL: InterVarsity Press, 1967), 83.

28 Gen. 4:25–5:32.

then the forefather of musical instruments and musicians followed. And then the forefather of both the Bronze Age and the Iron Age and the original inventor of all sharp metal tools appeared.

Listen carefully to the brothers' names—Jabal, Jubal, Tubal. That bit of their names that rhymes derives from the Hebrew word meaning "to produce."[29] They were producers. Their core identity was bound to an inherent inventiveness within them. Each son, whether by name or occupation, bore allusions back to Cain.[30] This is why Cain was preserved, to result in a "many-sided material culture developed among his scions."[31]

Our Tech Fathers

The fourth chapter of Genesis delivers a rapid introduction to our technological forefathers, made clear in three little phrases in the text: "the father of those who . . . ," "the father of all those who . . . ," "the forger of all." Each brother launched a whole industry of innovation that the author of Genesis wants us to trace to the present day. Man's technological progress is of biblical interest.[32] The brothers' influence tracks through history. From the Jubal of Genesis to the Jobs of Silicon Valley, every brand-new industry exerts a permanent and perpetual influence on the development of all subsequent human culture in the future.

But if we are to trace the influence of Cain's lineage to our lives today we are met with a big problem. These three inventors and their whole industries were soon to be washed away by a global

29 Kenneth A. Mathews, *Genesis 1–11:26*, vol. 1A, New American Commentary (Nashville, TN: Broadman & Holman, 1996), 287–88.

30 Cassuto, *Commentary on the Book of Genesis*, 235.

31 Cassuto, *Commentary on the Book of Genesis*, 230.

32 Herman Bavinck, *Reformed Ethics: Created, Fallen, and Converted Humanity*, ed. John Bolt et al., vol. 1 (Grand Rapids, MI: Baker Academic, 2019), 163.

flood. So if animal genetics, metallurgy, and the music industry owe their origins to the family legacy of three brothers whose progeny was entirely washed away in a global flood, how can they be said to be forefathers to these industries today?

The ark.

Stowed away in Noah's ship was God's chosen family, two creatures of every kind, and humanity's collective technological know-how. Noah may be ignored in the long history of technology, but he and his family are essential midwives to the unfolding story of human innovation. Noah was one of the most ambitious and brilliant of the ancient engineers, and he was God's chosen man to carry all the industries of Cain's lineage through the flood. By common language, Noah and his ark diffused all of mankind's preflood technological know-how into all the earth's post-flood population. That knowledge eventually manifested in human engineering feats of unbelief (Babel). But we must not rush past the ark too soon. In the modern marvel of his age, Noah carried the innovations from Cain's lineage and dispersed them as common knowledge into the new world. God's relationship to human innovation will not make sense if we fail to see these connections.

The chain goes like this. The ground rejected Cain. God rejected Cain. But God did not kill Cain. Instead, God chose to channel his common grace through Cain's lineage to bless the world. A murderous rebel and his rogue family became God's choice for unleashing new innovation into the world. Whole industries were carried by Noah into a new world and passed down to us today.

Human history begins in these two lineages. Abel's aborted lineage was replaced by a new son, Seth, and the story of God's covenant people continued. With Seth's arrival, the faithful began calling on *Yahweh*—"he who is who he is" and "he who causes to be what is." For the first time, the self-sufficient maker of the universe was now called on by his holy name.[33]

The Question

So where did the outpouring of ancient human innovation come from? Many people assume that elite technologists bubble up by a law of inevitability. One person in a hundred million will be a great inventor, and every age gets one or two of them, virtuosos like Archimedes, da Vinci, Edison, Franklin, Ford, Tesla, or Einstein. They discover new realities or assemble a bunch of cutting-edge things. They're rare.

Without saying it explicitly, the consensus is that inventors are simply the product of random chance or the result of cream rising to the top. But this idea is misguided. From what we learned in Isaiah 54, it's more accurate to say that Jabal, the forefather of nomadic herdsmen, was created by God and ordained for this purpose, to invent ranching and to understand animal breeding and to begin experimenting with what we now call genetics. And Jubal, the forefather of music and instruments, was created by God and ordained for this purpose, not simply to give amateurs something to play with, but to birth the industry of specialized musical professionals who master instruments for public celebration in bands and orchestras and the music industry.[34] And then Tubal-cain, the first cause of both the Bronze Age and the Iron Age, was created by God and ordained for this purpose, to be a blacksmith who would fill his community with sharp metal tools and weapons as "the forger of every *cutting* instrument of brass and iron."[35] He

---

33 Sarna, *Genesis*, 127.

34 Sarna, *Genesis*, 37.

35 Jewish Publication Society of America, Torah Nevi'im U-Khetuvim, the Holy Scriptures according to the Masoretic Text (Philadelphia, PA: Jewish Publication Society of America, 1917), Gen. 4:22; emphasis added.

hammered and filed down every sharp thing meant to cut: knives, sickles, and swords of war. And let's not forget the patriarch at the start of the story, Cain, the forefather of city developers, who was created by God and ordained for this purpose.

Just as soon as these innovators arise in the storyline of Genesis, no further explanation is given to us. The skill to build a city, breed animals, make music, and forge bronze and iron tools seemed to come out of nowhere. But behind the scenes, we know that they were ordained by God. Innovators are made by God.

But where do their innovations materially originate? That's the next question we must answer.

Ag Tech

To understand the material origin of our innovations we must follow another set of clues in the biblical text. And we get them in Isaiah 28:23–29, in the origins of agricultural technology. We begin in verse 23.

> 23 Give ear, and hear my voice;
> give attention, and hear my speech.

Isaiah says to us, "Stop whatever you're doing and listen up! Pay attention, because what you are about to hear is huge!" Okay, so what is he urgently trying to tell us? He's going to walk us through modern farming techniques (modern to Isaiah anyway).

First comes the preparation. Certain ground practices are necessary before a crop can be planted.

> 24 Does he who plows for sowing plow continually?
> Does he continually open and harrow his ground?

Does the farmer keep opening and closing the ground? No, he harrows the ground a few times. But who told the farmer to do this a few times and not perpetually? Who taught the farmer to follow this pattern for the best results? That's one open question.

Hold that question as Isaiah moves on to planting techniques.

> 25 When he has leveled its surface,
>   does he not scatter dill, sow cumin,
> and put in wheat in rows
>   and barley in its proper place,
>   and emmer as the border?

The field is smooth and tilled. The soil is loose. The ground is ready for seed. Now is the time to scatter dill and cumin seeds here and there. But harrow the ground for wheat; it grows best in straight rows. And on the borders, plant emmer. Each seed has an optimal *place* to be planted and a specific *way* to be planted.

So who taught the farmer how to multicrop? Who is the brain behind human planting techniques? Is it the farmer? No.

> 26 For he [the farmer] is rightly instructed;
>   his God teaches him.

Catch that. The master farmer learned his planting techniques directly from the Creator through his creation. Now hold on to that point for a moment as we move from planting techniques, skip over harvesting techniques, and study threshing tools and practices.

> 27 Dill is not threshed with a threshing sledge,
>   nor is a cart wheel rolled over cumin,

but dill is beaten out with a stick,
     and cumin with a rod.

28 Does one crush grain for bread?
     No, he does not thresh it forever;
 when he drives his cart wheel over it
     with his horses, he does not crush it.

Dill, cumin, grain: three crops for three purposes, each requiring
different tools and techniques. Dill is threshed with a wooden stick
and cumin with a metal rod. Grain gets driven over with a horse-
pulled cart. Note the escalating force; each technique is matched
to the crop to maximize the yield. The crop itself seems to teach
the farmer. Or does it?

As these verses were written, the first published volume of collected farming advice was still half a millennium away.[36] So who
gets the credit for teaching the farmer these threshing tools and
techniques?

29 This also comes from the LORD of hosts;
     he is wonderful in counsel
     and excellent in wisdom.

So who gets credit for matching the exact farming tool and
technique to each crop? Evolutionary progress? No. The ancient
farmer's guild? No. What about creation itself? No, not even creation. The Creator gets the glory, all of it. Why? Because the tilling
and planting and threshing techniques of the master farmer are

36 Marcus Cato, *De agri cultura* (160 BC).

all techniques learned from the all-wise and all-sufficient Creator. Contrary to the theory that stable farming practices led to the invention of the gods, it was God who invented stable farming practices.[37] God's eternal existence and his patterns in creation predate the farmer.

God's Providence over Us

As we will soon see, this text holds many consequences for our discussion of tech today. But first, we need to pause for one moment to speak again about God's people. This text lands in the middle of Isaiah because it's a key to unlock the whole book. God's relationship to the farmer is an important but secondary point. This text is about God farming Israel. And he knows what he's doing. He will not continue to till open his people's hearts with a scratch plow, and neither will he thresh them continually. He's the master agronomist, and he's moving his people toward a goal. He's husbanding history to bring about a spiritual harvest in his people. And if it feels like God is overfurrowing and overthreshing your life right now, he's not. His agricultural work in us, though painful, will yield a harvest of righteousness.[38] We are his vine. He is our vinedresser. He knows exactly how to bring about the optimal crop from your life. He is going to disc-harrow your life, prune your deadness, fertilize your roots, and watch over you during every season. God instructs the farmer because he is the master farmer of our lives. He has many tools and techniques he can use in us to bring about his goal. The work of God inside us is agricultural.[39]

37 Yuval Noah Harari, *Homo Deus: A Brief History of Tomorrow* (New York: HarperCollins, 2017), 90–91.
38 Heb. 12:11.
39 Matt. 13:1–23.

A Major Hurdle

So God teaches the farmer new agricultural techniques. But wait. Two objections spring up from this agricultural focus. The agrarian asks: "Doesn't Scripture teach us to dig no deeper into creation than the scratch plow, and to build no higher in the sky than a ziggurat?" And the technician asks: "This works great in agriculture—planting seeds, harvesting grain, listening to creation. But I work in a hermetically sealed, dust-free, clean room with a mask and white Tyvek coveralls. I work with cobalt, indium, tantalum, robots, and software. God's voice is mute where I work. I don't hear creation. Isaiah 28 may work in agrarian contexts, but not inside a semiconductor fabrication facility. Isn't there an insurmountable difference between a primitive-basic tool, like a garden hoe, and tech today, like a smartphone or a nuclear power plant?"

I hear both questions, and both are important. So, I pray: God, show me the relevance of Isaiah for the digital age. In seeking to understand the relationship between the family farm and SpaceX rockets, here are three responses to consider: (1) farming is our primary tech; (2) all tech has ancestors; and (3) every innovation is patterned by agriculture.

Farming Is Our Primary Tech

First, farming is our primary tech. In fact, farming is likely "the most fundamental of all human technological breakthroughs," the base technology behind all other technologies. Like the green, plastic base plate on which most of us stacked our first LEGO houses, farming is the technological base that other technologies build upon. Humanity's first technological revolution was agricultural, and only once "food production technology had taken hold, a chain reaction of other technological developments followed at an ever-quickening speed. The transformation in food production was followed by the development of metallurgy, the invention of the wheel, the perfection of systems for writing and recording information, and other technical innovations which had powerful effects on human culture."[40]

When you gather your food here and there, you need a search party full of foragers. But when you grow your food in acres of crops next door, the gatherers can fill their time with other pursuits. Sedentary crop cultivation brought major social changes and supported larger communities that could live in one location. That made it possible to own more things, possess more tools, and live behind protective city gates. Farming made the city possible, and the city helped to free people from food concerns. Farming led to cities, and cities produced innovators and inventions—all because farmers could feed more people than themselves. (Today, for example, one American farmer can feed about 160 people—who, no longer burdened with the daily concern for raising their own food, can fill other roles in society.)

Food sourcing provided a stable base for other technologies to develop. Without farming technology, we'd all be called on to forage, hunt, and gather to sustain us and our small communities. Agriculture is a primary technology; it makes all other technological progress possible. Without the sickle and the plow, without primitive farming and its tools, there would be no cathedrals and "no European voyages of discovery."[41] Agri-business is a primary development, and new possibilities of human discovery stand on the shoulders of farmers.

---

40 Harvey Russell Bernard and Pertti J. Pelto, *Technology and Social Change* (New York: Macmillan, 1972), 317–18.

41 Václav Smil, *Energy and Civilization: A History* (Cambridge, MA: MIT Press, 2018), 52–53.

All Tech Has Ancestors

Second, all our tech has ancestors. Farming itself is a prime example of technological advance and illustrates how primitive technologies adapt, modify, merge, improve, and become more powerful and complex technologies later. Technology theorist W. Brian Arthur calls this "combinatorial evolution." All technologies are blends of previous technologies or "fresh combinations of what already exist," he says. It's like the techno-compilation of the iPhone. Micro-components, like cameras, voice recorders, touch screens, music players, and speakers, each of them technologies themselves, combined into one new macrotechnology. The recursive process continues indefinitely as simple tools merge into new technologies, which become a network of possibilities, into which even more complex technologies emerge in the future. In a sense, technological progress is organic, as if "technology creates itself from itself." Technology forms a "rich interlinked ancestry" that grows more complex and sophisticated over time. As the heredity grows, our machines of the future will think for themselves and act for themselves, becoming "self-configuring, self-optimizing, . . . self-assembling, self-healing, and self-protecting." They resemble a living organism of sorts. "But it is only living in the sense that a coral reef is living."[42]

So the interlinked ancestry of the iPhone can be traced back to more primitive innovations. Think of the glass company, Corning, that first mastered kitchen glass a century before mastering Gorilla Glass, now used on every iPhone. The story of your iPhone screen is more than a century old. And that's just a sliver to the full story. If you've ever played a video game like Sid Meier's *Civilization*, you know this. You,

---

42 W. Brian Arthur, *The Nature of Technology: What It Is and How It Evolves* (New York: Penguin, 2009), 18–24, 189, 207.

as the ruler, build a civilization based off an ever-evolving technology tree. One primitive technology leads to the discovery of more advanced innovations, as centuries roll on from 4000 BC to AD 2100.

Take language. Language is a base innovation, gifted to man by God from the beginning of time and then multiplied at Babel. Language made it possible for man to talk with God and with other humans. It made it possible for the first humans to name animals, even sometimes to speak to animals.[43] It made it possible to name stars and name children, to tell stories and build cities. Language formed hieroglyphics and then alphabets, writing, books, printing presses, libraries, telegraphs, texts, tweets, and the digital code of programmers. The gift of language also meant that God could reveal himself in incredibly detailed and intricate ways that would endure for millennia. Language carries God's will and intent for creation, discloses the life-giving message of the gospel, and serves as a channel to renew the inner man.[44] Language represents the oldest and fullest and richest of the tech trees.[45] But the ancient alphabet begat more than just words—oral stories led to written alphabets, which led to libraries filled with scrolls, which led to written laws, which led to democratic thinking, which led to more accessible printing techniques, which led to the bound codex book, which led to widespread literacy—and on and on until you arrive at the Internet. Without ancient alphabets, you don't get the Internet. It's all one, long, ancestral line, connected with rudimentary beginnings.

The same steady growth appears in agricultural technology, year after year, generation after generation, like thick tree circles mark-

---

43 Gen. 3:1–24.
44 Rom. 10:17; Col. 3:10; Heb. 4:12.
45 A good theme in John Dyer's very good book *From the Garden to the City: The Redeeming and Corrupting Power of Technology* (Grand Rapids, MI: Kregel, 2011); see esp. pp. 51–54.

ing seasons of abundance. One generation of advances calls forth more aggressive advances in the future. For example, the earth has enough organic nitrates in waste and compost to grow food for about three billion people. That's it. So how can this planet provide food for four billion, seven billion, or ten billion people?

The answer is bound to the following question. What was the single, most important invention of the twentieth century? The question was put to a TED audience by scientist Václav Smil in 2000, when the global population stood at six billion. Answers from the crowd were shouted out: penicillin, air-conditioning, radios, televisions, computers, the Internet, human flight, and nuclear power. "You are all wrong," Smil announced from the stage. "We have six billion people on this planet; half of them would not be here without that invention which you haven't named—the Haber synthesis of ammonia." Due to the limits of organic nitrogen in the global ecosystem, without ammonia nitrogen fertilizers, "half of the population of this planet wouldn't be here. There is no other technique, no other invention, without which today there wouldn't be half of the people here. So, by far, [this is] the most important invention [of the twentieth century]. And most people are not even aware of it."[46] If scientists are right and the earth can support a mere forty million hunter-gatherers, or three billion city dwellers through organic farming, then man-made ammonia prevents most of us from starving today.[47]

So given the earth's apparent lack of organic nitrogen, who teaches us how to feed four and eight and ten billion people on the

46 For the full story and the science see Václav Smil, *Enriching the Earth: Fritz Haber, Carl Bosch, and the Transformation of World Food Production* (Cambridge, MA: MIT Press, 2004).
47 Carl Sagan, *Pale Blue Dot: A Vision of the Human Future in Space* (New York: Ballantine, 1997), 316.

planet? God does. The Creator teaches us how to scale agriculture to meet human demand. Agricultural technology is dynamically progressive and intentionally builds off the primitive discoveries of Isaiah 28. Every tech sprouts from this ancestral, agricultural root.

Every Innovation Is Patterned by Agriculture

Third, every innovation is patterned by agriculture. Another way to say it is that every material innovation is equally rooted in creation—not only primitive agricultural technologies but also modern technological innovation. John Calvin makes this point well. First, whether we are talking about "the most untutored and ignorant persons," or about farmers or doctors, or scientists, or astrophysicists, God's wisdom is taught to all. Whether you never went to college or have a PhD in botany, whether you grow one pot of tomatoes or harvest a million bushels of soybeans, Calvin says, "It is clear that there is no one to whom the Lord does not abundantly show his wisdom."[48] Isaiah 28 reveals an exponentially larger principle at play in the world. God teaches us all, and liberally so. The spiritual health of Isaiah's farmer is irrelevant. Every thinking human being—believer, unbeliever, skeptic, agnostic, atheist—has been doused with divine wisdom, directly from God, for how to engage the created order.

A scientist in a lab coat is not more removed from the Creator than a farmer in a field. In fact, Calvin writes that scientists are actually plunged "more deeply into the secrets of the divine wisdom." Isaiah's farmer is a simplified version of an elastic paradigm that stretches from the ag co-op to NASA's mission operations control room, as Calvin explains in his footnote to Isaiah 28:26:

48 John Calvin, *Institutes of the Christian Religion*, ed. John T. McNeill, trans. Ford Lewis Battles (Louisville, KY: Westminster John Knox, 2011), 1.5.2.

A passing observation may be made, and indeed ought to be made, that not only agriculture but likewise all the arts which contribute to the advantage of mankind, are the gifts of God. All that belongs to skillful invention has been imparted by him to the minds of men. Men have no right to be proud on this account, or to claim to themselves the praise of invention, as we see that the ancients did, who, out of their ingratitude to God, ranked in the number of the gods those whom they considered to be the authors of any ingenious contrivance. Hence arose deification and that prodigious multitude of gods which the heathens framed in their own fancy. Hence arose the great Ceres [the god of agriculture], and Triptolemus [the god of sowing and milling grain], and Mercury [the god of importing and exporting], and innumerable [other gods], celebrated by human tongues and by human writings. The prophet [Isaiah] shows that such arts ought to be ascribed to God [not gods], from whom they have been received, who alone is the inventor and teacher of them. If we ought to form such an opinion about agriculture and mechanics, what shall we think of the learned and exalted sciences, such as Medicine, Law, Astronomy, Geometry, Logic, and such like? Shall we not much more consider them to have proceeded from God? Shall we not in them also behold and acknowledge his goodness, that his praise and glory may be celebrated both in the smallest and in the greatest affairs?[49]

---

49 John Calvin, *Commentary on the Book of the Prophet Isaiah*, trans. William Pringle (Edinburgh: Calvin Translation Society, 1853), 2:306; quote slightly modified for readability. So also Charnock, who writes: "The art of husbandry is a fruit of divine teaching (Isa. 28:24–25). If those lower kinds of knowledge, that are common to all nations, and easily learned by all, are discoveries of divine wisdom, much more the nobler sciences, intellectual, and political wisdom." Charnock, *Complete Works*, 2:20.

For centuries, pagan cultures have impulsively credited the gods for human innovation. Calvin says that this impulse is wrongly misdirected away from the one, true God. God alone is the Creator of human innovators (as Isaiah 54 taught us).

More importantly, notice how Calvin's logic moves freely from the rudimentary farming tools of the ancient farmer to more complex machines, then to science, medicine, law, and physics. My first reading of this Calvin quote set in motion what would become a Copernican Revolution in my own understanding of God, creation, agriculture, science, and engineering—how our modern technologies all orbit in the same system. There's no need to segment basic agricultural technique from more complex machines or from medical, electrical, and even genetic technologies. Science, like agriculture, is the art of listening to the Creator, the art of following out the patterns and possibilities that God coded into creation.

Man's dominion of the earth "includes not only the most ancient callings of men, such as hunting and fishing, agriculture and stock-raising, but also trade and commerce, finance and credit, the exploitation of mines and mountains, and science and art."[50] Isaiah's logic of the farmer carries on to electrical and digital sciences and quantum computing. It carries over into the beauty and power of simple equations like $E=mc^2$ or Euler's equation. It carries over into rocket science. Konstantin Tsiolkovsky published in 1903 what would become the Tsiolkovsky rocket equation to calculate the velocity of rockets. It carries over into flight science. Daniel Bernoulli published in 1738 what would become the Bernoulli effect to help explain airplane lift. God taught us how to build rockets and planes like he taught the ancient farmer to grow crops. In every human

50 Herman Bavinck, *The Wonderful Works of God* (Glenside, PA: Westminster Seminary Press, 2019), 189.

discovery we find the Creator's instruction. God is our tutor, and he ordains every link in the chain of technological revolution.

Over time a unity to all our scientific discovery begins to emerge. We begin to see the design of a single architect. Scientific endeavor is like humanity building a stone temple, writes Kuyper. "The entire temple is constructed without a human blueprint and without human agreement. It seems to arise by itself," he said. "Each one quarries his own little stone and brings it forward to have it cemented into the building. Then comes another person who removes that stone, refashions it, and lays it differently. Working separately from one another, without any mutual agreement and without the least bit of direction from other people, with everybody milling about, everyone going his own way, each person constructs science as he thinks right." Over centuries, as incremental improvements change portions of the building, "out of this apparently confused labor, a temple emerges, displaying the stability of architecture, manifesting style," all because there is an "Architect and Artisan whom no one saw." None of it emerged by accident. Science follows a plan laid down by God. He patterned the end of creation and then gave creation innovators with creative imagination. Science is God's invention. "This means nothing else except to say and to confess with gratitude that God himself called science into being as his creature, and accordingly that science occupies its own independent place in our human life."[51] Science takes on a unified life of its own.

This illustration from Kuyper, of science as a stone temple built over time, supports what John Calvin taught centuries before. Calvin lived long before SpaceX or Apollo 11, but he was already tracing the trajectory from primitive building tools to advanced

---

51 Abraham Kuyper, *Wisdom and Wonder: Common Grace in Science and Art* (Bellingham, WA: Lexham Press, 2015), 45–46.

aerospace physics. He was already grafting them together into one God-ordained tech tree, arguing from lesser tools to more potent technologies and more refined scientific discoveries. If the farmer made a stick to beat cumin seeds, later someone would take the functions necessary to farm, combine them, and automate them into a steam-powered mill. This process continues today. Every day we hear new possibilities from the Creator. He never stops teaching, so we have no excuse to stop listening for his instruction or correction. Each brick in the temple continues to be reshaped and improved. Various old innovations merge into new, better innovations of the future.

But all of this new innovation does not mean our past innovations disappear. On the contrary, when Kevin Kelly investigated old innovations, from stone hammers to steam engine valves, he noticed that every human innovation that has ever been adopted is still in production somewhere.[52] We may think of innovations as expiring and getting replaced, but the most useful ones are permanent. And if all innovations remain relevant, perhaps God is right now teaching small, remote communities to self-sustain on foods raised with primitive hand tools. New tech doesn't scrap old tech.

But as people pack tightly together in cities, those cities will depend on colossal farming operations where food can be grown and processed outside of the city and shipped in. Industry at this scale will always be imperfect. But the city calls for escalating advances, so one era of farmers would listen to the Creator and use rods and carts. A later generation would listen to the Creator and use paddle-wheel irrigation systems and sickles. Another generation would listen to the Creator and use irrigation sprinklers, self-driving combines,

---

52   Kevin Kelly, *What Technology Wants* (New York: Penguin, 2011), 53–56.

and synthesized ammonia to boost nitrogen levels in the soil. These incremental improvements are how God accommodates this planet to host larger and larger populations. It is the pure generosity of God that causes the sunshine and rain, not to mention synthetic nitrogen, to fall on the farmlands of his enemies.[53]

So while the Creator teaches communes in Puerto Rico how to grow native varieties of maize, he can simultaneously teach industrial farmers in Nebraska how to grow over a billion bushels of corn. God continues to instruct farmers every day.

Catechism of Experiments

At whatever scale, these ongoing dynamics of innovation work because creation is a catechism we read by science and experiments. God speaks through human investigation. "Did the first farmer receive a manual?" asks Kuyper. "Did God send him an angel to demonstrate everything? Did God give him an oral revelation? None of these things. God gave him the soil, a head to think with, hands to work with, and (besides these) a basic *hunger*. God stimulated him by means of this drive. God taught him to think about things. And thus he had to try things. First one thing, and when that did not work, something else, until finally one person found this and the other that, with the results confirming that this was the right solution."[54]

Basic hungers and base needs eventually give way to more complex hungers and more debatable "needs" in prosperous societies.[55] But the same forces are at play. All technological change is driven

53 Matt. 5:45.
54 Abraham Kuyper, *Common Grace: God's Gifts for a Fallen World* (Bellingham, WA: Lexham Press, 2020), 2:585.
55 Arthur, *Nature of Technology*, 174–75.

forward by internal human urges, which drive new exploration into the possibilities within the patterns of creation. Farming technology advances because of hunger. Medical technology advances because of illness. Our technologies can also advance through less virtuous hungers for comfort and wealth and fame and power. All sorts of internal desires drive mankind to experiment. But no matter the motive, innovation can only follow along a pattern that God himself coded into the created order.

Within the created order, God limits all our possibilities. It's a myth to assume that all human innovation will advance indefinitely. Innovations often hit ceilings, like in the speed of commercial airlines. We learned how to swirl category-five hurricane force winds out the back of jet engines, but at 575 mph, jets today cruise at about the same speed as they did sixty years ago. Our flights are more safe and fuel efficient, but the optimal cruising speed hasn't changed in six decades. The creation seems to have capped the optimal cruising speed of commercial jets. Man and machine optimized to God's pattern of creation.

Again, says Kuyper: "Through endless types of experiments God has taught us all that we now know, and through all kinds of experiments our knowledge continues to be enriched."[56] As we experiment with creation, as we seek new ways of doing things, the Creator teaches us. "God does not give the farmer a special revelation as to how he must plow, sow, harrow, weed, and thresh," writes Kuyper. "The farmer must learn by trying. But as he constantly experiments with new ways and when through much trial and error he has learned the hard way, then it is true what the prophet Isaiah says about this process: 'His God teaches him.'"[57] At our best, "we

56 Kuyper, *Common Grace*, 2:585–86.
57 Kuyper, *Common Grace*, 2:616.

allow ourselves to be instructed in more and better ways by our God—not only in the Heidelberg Catechism [spiritually] but also in the catechism of agriculture, in the catechism of industry, and in the catechism of commerce. For it is obvious that everything we have said about farming on the basis of Isaiah 28 is just as applicable to all the rest of human endeavor."[58] If you extrapolate the dynamics that drive farming technology, you will see those same factors at play in all human endeavors.

Atheists claim that Scripture contains all that God has to say or to teach us.[59] But this is naïve. God also teaches us through the catechism of science. The scientist becomes "the priest of creation," writes Torrance, "whose office it is to interpret the books of nature written by the finger of God, to unravel the universe in its marvelous patterns and symmetries, and to bring it all into orderly articulation in such a way that it fulfills its proper end as the vast theater of glory in which the Creator is worshipped and hymned and praised by his creatures."[60]

Even our attempts to stop viruses should bring him praise.

Battling Coronavirus

In March 2020 I brought these ideas about technology to Seattle, in a gathering overlooking the downtown stadiums on the bay. From a fourteenth-floor conference room, I watched a team of men painting the football field lines for the weekend's XFL game. The game never happened. The entire season was canceled a few hours later due to the coronavirus pandemic. Early in the outbreak,

---

58 Kuyper, *Common Grace*, 2:585–86.
59 See, e.g., Harari, *Homo Deus*, 213.
60 Thomas Forsyth Torrance, *The Ground and Grammar of Theology* (Charlottesville, VA: University Press of Virginia, 1980), 5–6.

Seattle was America's epicenter. And as the city shut down and the streets emptied, it was also a remarkable moment to address the pandemic's spread from Southeast Asia to North America. We watched this virus move from country to country, state to state, marching toward a global pandemic and all before our eyes, live, on social media. As the virus spread, a genome database continued to grow too. Scientists across the globe tracked the RNA structure of the virus as it mutated and changed from person to person. These mutations made it possible to build a transmission chain, to trace infections back to earlier sources, even back to a likely original host. It is stunning that we have the technology and open platforms to track in real time the thumbprint of a virus as it morphs and spreads across the globe.

Also amazing were the ways that medical researchers brain-stormed and collaborated to combat the virus. Who taught us to track and stop viruses? Do nineteenth-century germ theorists get all the glory? No. Do twenty-first-century vaccinologists who de-signed mRNA vaccines get all the credit? No. God gets the credit and the glory for every innovation that heals. To paraphrase Isaiah's punchline from the divine origin of ag tech, our latest discoveries into infectious diseases also come from the Lord of hosts, who is wonderful in counsel and excellent in wisdom.[61]

But the generosity of God can be hard to see when we put so much of our confidence in science and human innovation. In April 2020, when the first wave of coronavirus cases mercifully began to decline in New York City, Governor Andrew Cuomo took to televi-sion to celebrate his administration's work and the hygiene of New Yorkers. "The number is down because *we* brought the number

---

61 Isa. 28:29.

down. God did not do that."[62] The implication is that whenever some good thing happens in the world that has no scientific explanation, God gets the credit for it. Fine, if that's what you want—says the secularist—give God credit for whatever cannot be explained. But what *we* accomplish, scientifically and medically, by figuring out how to slow and stop viruses, all credit goes to the emphatic, italicized, *we*. *We* did that. *We* brought the numbers down. Now, I'm not picking on Cuomo. He's just a recent example of the natural sentiment of a man with little awareness of God, unaware that the only reason we can fight a virus is that the Creator is actively speaking and teaching us how to follow creational patterns. God is at work instructing us inside laboratories, through experiments, and within clinical trials. The felt urgencies that drive the science include trying to save lives, salvage the economy, and reopen local economies. But by our urgency God is teaching us to combat infectious diseases.[63]

A few months later, recovering from the virus in a hospital, President Trump took to video to praise new experimental drugs he'd been given. "Frankly they're miracles, if you want to know the truth. They're miracles. People criticize me when I say that. But we have things happening that look like they're miracles coming down from God."[64] That's solid theology. The cutting-edge products of science are divine gifts.

62 "Governor Andrew Cuomo New York Coronavirus Briefing Transcript," rev.com (Apr. 13, 2020).

63 "All disasters that threaten or occur come from God, in the same way that every way, hidden in nature, to prevent disasters does as well. . . . We must no longer withdraw half of life from God's honor. We must not imagine that science could do anything on its own, as if science in this way had forged a weapon to defy God or if it had somehow succeeded in rendering God's lightning powerless. Science has no power whatsoever, and there exists no force in nature that is not worked by God." Kuyper, *Common Grace*, 2:597.

64 "President Donald Trump provides update on health, says he'll 'beat coronavirus soundly,'" *Global News*, youtube.com (Oct. 3, 2020).

When coronavirus vaccines began shipping in America in December of 2020, just nine short months after the pandemic first spiked in America, Dr. Kathrin Jansen, a scientist and the head of vaccine research for Pfizer, was asked if the rapid new vaccine was a miracle. "We can call it a miracle," she said. "But a miracle always has a sense of, it just happened. It didn't just happen. Right? It was something that was deliberate. It was done with passion. Urgency. It was always having in your sight that devastating disease."[65] Yes. But perhaps we can preserve room for scientifically discovered miracles? Or at least preserve a place for scientifically discovered gifts from God?

With vaccinations shipping and controversy swirling about the safety of the rapidly developed injections, the staff of the National Institutes of Health gathered to be vaccinated on live television. One of the last to roll up his sleeve was NIH Director Dr. Francis Collins, a Christian. A moment after his injection, he moved to a podium to call the historically fast vaccine a light at the end of a long dark tunnel called COVID-19, "a light made possible by the power of NIH science and our many partners." Like Pfizer. He then closed the gathering with a prayer of gratitude to God by reciting Psalm 103:2–5, the famous "Bless the Lord" psalm, with particular emphasis on the God who "heals all your diseases," and "satisfies you with good so that your youth is renewed like the eagle's."[66] Scientific innovation is nothing short of a divine gift.

Whether to the farmer in his field or the epidemiologist in his laboratory, the Creator instructs us through the patterns and pow-

65 "How the Pfizer-BioNTech COVID-19 Vaccine Was Developed," *60 Minutes*, youtube. com (Dec. 20, 2020).

66 "LIVE: Anthony Fauci, Alex Azar Receive Covid-19 Vaccine in Washington, D.C." *Bloomberg Quicktake: Now*, youtube.com (Dec. 22, 2020).

ers he intentionally set in place in his creation. This is true of genes and germs and geysers. Long before humans imagined enclosing steam to power the nineteenth century, geysers of powerful steam erupted from the earth. And God set prevailing winds across the globe—predictable airstreams and harnessable power to move huge wooden boats across waterways to ship merchandise and explore lands.[67] Or think of lightning bolts, crackling from the sky at a rate of about one hundred per second. God handcrafts each one.[68] Each bolt cracks with a declaration: "You can use this to power your cities!" So Ben Franklin asked himself, "Hmm, I wonder if lightning is electricity?" And then he mythically caught one bolt with a kite in 1752 and proved it. Yes, lightning is electricity, and by the bolt the Creator taught whole generations to make light bulbs, to electrify cities, and to inaugurate the computer age and the smartphone age and the age of the electric car. Or think of the space-age scientists who aspire to harness nuclear fusion to power our cities. Is nuclear fusion inspired by CGI effects in a sci-fi movie? No. As we will see later, this supreme form of power was God's idea in the burning of our sun (and innumerable stars bigger than our sun).

Many Christian critiques of technology assume that all human innovation is an inorganic imposition forced on the created order. I suggest an opposite way of thinking. The priests of creation, the scientists, uncover patterns in creation. Then the deacons of discovery, the technologists, exploit those patterns for human flourishing. All human innovation in that sense is organic. Apart from the possibilities of creation, human innovation is impossible. For any

---

67 William Bates, *The Whole Works of the Rev. William Bates* (Harrisonburg, VA: Sprinkle, 1990), 3:78–79.

68 Ps. 135:7.

drug, vaccine, or technology to work, it has to follow a pattern. We don't make the patterns. We can only follow them as we discover new divine gifts.

Balrogs

When Carl Sagan attempted to prove God's hostility to human science, he did it by claiming that Adam and Eve were forbidden from the tree of "knowledge and understanding."[69] He wrongly assumed that the forbidden tree in Eden was God's attempt to keep man forever separated from scientific learning and technological prowess. Many Christians slip into this error too, assuming that God is threatened by man's scientific awareness. The faulty logic thinks that if we continue innovating, we are sure to unearth something forbidden to bring down the curse of God upon us all.

Or to use a mythic mining metaphor, if we keep digging deeper into this world to find new innovations, we will eventually unearth some new nefarious power, like covetous dwarves who dug too deep into the earth seeking mithril, a metal worth ten times the value of gold. Those dwarves, warned the wizard Gandalf, "delved too greedily and too deep, and disturbed that from which they fled, Durin's Bane."[70] Greedy dwarves awakened a nightmare, a monstrous evil, a balrog. Many Christians, too, are technologically timid, fearful of awakening a balrog, as if we could violate the created order and unearth some evil power we were never meant to discover. On the contrary, if God did not want us to discover something—raw materials or natural laws or potential powers—he simply didn't code it into the pattern of his creation. God sets the height and width and depth for the sandbox of our discovery.

69 Sagan, *Pale Blue Dot*, 53.
70 J. R. R. Tolkien, *The Lord of the Rings* (New York: Mariner, 2012), 317.

This goes back to the Creator's intent. God filled this globe with a precise and intentional distribution of minerals and metals and liquids and gases and atmospheric pressures and gravity and oxygen and water to produce the world that we have currently. Every new discovery by God's image bearers illuminates the mind of the Creator in his design of creation. But it also spotlights the apex of his creation: the image bearers themselves.

Creation exists because God spoke. No other cause beyond his intentional pattern can explain the microscopic properties of quarks or the expansive volumes of mass. God alone measured out the distances in space and portioned out the volume of water and materials needed on our planet.[71] God calculated every square inch of creation with perfect balance and precise patterns and set it in place for us to investigate and cultivate. This planet is full of surprises for us. But not for him.

LEGOs

Here's a simple illustration. Imagine a fifty-five-gallon drum filled with sixty thousand LEGO pieces of every size, shape, and color. From that drum we could build anything we wanted, a bunch of small things or one big thing. We could use some of the pieces, most of the pieces, or all of the pieces. Then we could take the pieces apart and rebuild something different. Now to finite minds like ours, that drum represents an infinite number of combinations that we could imagine, build, and rebuild. The infinite possibilities are what make LEGO bricks so captivating to the human imagination. We could never calculate in our own heads every possible outcome.

71 Isa. 40:12.

But to an infinite mind, that drum of LEGOs has a very finite number of outcomes, all of them predictable. God's relationship to creation is not finite, as if he gives us a drum of pieces that we can mix and match and build things he never could have imagined. He is infinite, and he already knows every combination, permutation, and limitation of what we can make. He not only knows the possibilities, but he also set the boundaries of those possibilities. He knows the realities we have discovered. He patterned the possibilities we cannot yet imagine. He selected the exact number, size, concentration, and color of every piece. Whatever concoctions we create all conform to the limits of the pieces he first made available to us. And these patterns not only apply to the bricks but also to how the pieces connect and hold together in the face of natural laws like gravity.

The Creator determines what should be made, codes those possibilities into creation, and leads his image bearers in a dynamic process of discovery, innovation, and improvement throughout the history of man. He already invented every LEGO blend and then threw all of those pieces into one drum. To a finite mind, these possibilities seem infinite and unlimited. To an infinite mind, the possibilities are finite and intentionally limited.

The LEGO illustration is a very simplified explanation of something far more complex. The drum of creation contains the entire material composition of the earth and our universe, and with it, every latent possibility guided by every natural law. God coded every pattern and boundary into creation. Our inevitable innovations are bounded within the potentialities of the created order. The Creator is free to make things from nothing. We cannot. We invent within the strict boundaries of raw materials and natural laws, boundaries set in place by the Creator himself. God patterns creation, down to each and every element on the periodic table of elements and its relative volume and availability to us. God alone sets in place all the patterns of human discovery we follow. He sets around us limits that, quite frankly, to a finite mind, feel like no limits. But the limits are there, and they are inviolable, because we have no access to what God does not make available to us.

From Nothing

So we return to what it means for God to make all things from nothing. God commands, and all things appear (Ps. 148:5). He simply "calls into existence the things that do not exist" (Rom. 4:17). So it is "by faith we understand that the universe was created by the word of God, so that what is seen was not made out of things that are visible" (Heb. 11:3). Thus the writer of Revelation, in speaking to God, says that God is worthy of all glory because he "created all things, and by your will they existed and were created" (Rev. 4:11). Creation exists for one fundamental reason: God was willing. Christ is the Creator, because on the one hand, "all things were made through him," and on the other hand, "without him was not any thing made that was made" (John 1:3). Get that double negative. Apart from Christ there is nothing made that was made. In Christ, "all things were created, in heaven and on earth, visible and invisible, whether thrones or dominions or rulers or authorities—all things were created through him and for him" (Col. 1:16).

Creatio ex nihilo means that the creation is secure and safeguarded by God, not by itself. No human activity can undo creation's ultimate stability because, right now, Christ "upholds the universe by the word of his power" (Heb. 1:3). His global upholding means that "since the universe is not only created out of nothing but maintained in its creaturely being through the constant interaction of God with it," the entire material universe "is given a stability beyond anything of which it is capable in its own contingent state."[72]

Christ upholds the stability of the universe by the word of his power. Christ is the origin, the sustainer, and the ultimate goal of creation. He has not rejected his sinful creation; he has redeemed it. Christ, the Creator, restores meaning and purpose to the material universe in three other important ways. By incarnation Christ entered into his material creation to confirm its value and meaning. By crucifixion Christ put an end to the tyranny of sin and vanity over his creation. By resurrection Christ inaugurated the new creation, restoring the destiny of his material creation, of our bodies and the cosmos itself. From within the material creation itself, Christ reaffirms its value, sets it free, and inaugurates its final telos. In other words, "the chief basis of human science, technology, craft, and art is therefore christological."[73]

Exploring all the implications of the Christ-centered basis of technology would require another book. Here I can simply state the glorious fact. Apart from Christ there is no art, no science, no technology, no agriculture, no microprocessor, and no medical innovation. Apart from Christ, we would have no iPhones. Nothing that now exists, visible or invisible, can exist if it first didn't exist in the mind of the Creator. This covers every single thing that's visible (like rare earth elements), invisible (like the soul), and nearly invisible (like atoms, RNA, and DNA). It includes all primary things, like the moon and oceans, and all derivative things, like farming techniques and smartphone apps. Apart from Christ there is nothing. The value of all things is relative to him.

72 Torrance, *Divine and Contingent Order*, 21.
73 Gunton, *Christ and Creation*, 123–24.

God learns nothing. He discovers nothing. He does not investigate like a scientist or experiment like a biochemist. He has no R&D lab. He needs no trials or errors. As Creator, God already knows everything about his creation, and he knows all things about his creation because he is the cause and genesis of all things inside it. God is the genius behind farming, not because he is the oldest observer of human farming tricks. He knows how farming works because the entire process for grain—from tilling to sowing to watering to harvesting to baking—was patterned by him and within him. When God spoke that pattern into being, we ended up with soil, oxygen, nitrogen, rainwater, and sunshine. He gives us our daily bread through a pattern of life he instilled into the creation. Creation brings forth these gifts.[74]

So don't pulverize bread grains. Don't thresh them forever. Run the cart wheel over these grains with a horse, but don't crush them. And don't gather dill with a threshing sledge. Dill is beaten out with a stick, and cumin with a rod. So the dill farmer whittles a stick, and God is not observing and nodding and thinking, "Oh, interesting. I didn't see that coming." No, even the farmer's techniques and tools were coded into the created order by the Creator himself. The farmer wields the stick, and the only right way to speak of that stick is that this technique and process was learned directly from the Creator himself, from his wonderful counsel and excellent wisdom. It's an art to be learned. Farming sticks, rods, wheels, axles, levers, cranks, toothed gears, and pulleys existed in the mind of God before the world was ever created, because the Creator patterned all of them in himself before the first seed sprouted.

74 Gen. 1:11.

By his intentional patterns in creation God called forth science and technology. Science discovers the patterns. Technology exploits the patterns. God is the genesis of all human technology. That is true for ancient shovels and for the combines and tractors used today. God patterned primitive tools, and he patterned more advanced technologies to produce food. God filled the LEGO drum, and we play with the possibilities. *Play* is the right word. Science and technology are not abstract fields that self-operate; they are the product of image bearers engaging the material universe with reason, imagination, and creativity. The playfulness of creation echoes in the playfulness of human discovery. So whatever technologies work to bring flourishing to this planet and to our lives existed first in the mind of God—long before they became a discovery of the inventor, who normally gets all the glory and the cash.[75]

In light of creation's design, I am not surprised that Einstein broke out in "rapturous amazement at the harmony of natural law, which reveals an intelligence of such superiority that, compared with it, all the systematic thinking and acting of human beings is an utterly insignificant reflection."[76] The Creator is an all-wise being who intentionally filled creation with natural forces and materials, intentionally designed a whole realm for man's learning, exploration, and wonder. The design is so intelligent that no human invention surprises the designer. And the design is so intentional that every pattern fulfills the Creator's eternal designs. So Calvin is free to call God the "lone inventor" because he's already designed

75 Petrus Van Mastricht, *Theoretical and Practical Theology* (Grand Rapids, MI: Reformation Heritage, 2019), 2:251–291.

76 Albert Einstein, *The World as I See It*, trans. Alan Harris (London: Bodley Head, 1935), 28.

in himself every technological possibility that can be imagined or produced in his creation. No technological possibility, no mix of raw materials or synthetic materials, can surprise God. He not only created the countless possibilities of human innovation; he also created every potential inevitability long before the inventor's own "aha" moment. And if every conceivable invention we could imagine was already patterned within the mind of God from eternity past, every single one of our discoveries should cause us to marvel at the limitless depth of who he is.[77]

Today, few farmers thresh grain with a handheld flail on a threshing floor. Instead, giant, green John Deere combines reap, thresh, and winnow in one process, guided by GPS coordinates, pulled along by more than 500 horsepower, and connected to an automatic wireless support network. Our ever-changing technologies are like signs at the end of a row of corn that advertise the seed that is planted there. The signs don't point to the farmer, and the technologies don't point to the inventors. All tech points to the inventor of all inventors, the one who made all things from nothing, and who prepatterned the boundaries of every one of our discoveries up until now and into the future. We innovate because he continues to speak and teach us today. Thus, the ultimate point of technology (in any age) is to point us back to the glory and the generosity and the majesty and self-sufficiency of the Creator himself. And the ultimate goal of technology is to usher us deeper into the creative genius of God, to direct our hearts to God, to adore him and to thank him for our daily bread. God's glory is the end of creation and the aim of all our innovations. He is worthy of our lives, worthy of our best inventions, worthy of all praise.

77 Van Mastricht, *Theoretical and Practical Theology*, 273.

Waiting

Despite bigger yields per bushel or better bread for baking, we don't adopt every technological possibility into our lives. And the people of God don't venerate the city dwellers of Jabal, Jubal, and Tubal-cain either. We celebrate the early city-avoiding men: Abraham, Isaac, and Jacob. In the Bible, cities are where people aggregate and proliferate technologies in order to buffer themselves from God.[78] Today, cities are epicenters of technological power, granting each of its citizens a growing illusion that they control the world. Techno-mastery is the atmosphere of the city, and it breeds atheism. It is only from inside a city that man can "with impunity declare himself master of nature. It is only in an urban civilization that man has the metaphysical possibility of saying, 'I killed God.'"[79] So God's faithful children have often avoided cities.

Much more must be said about cities later. But Babel is a prototype for what cities do best: they create altars for man to worship man. And yet after ending this one city, God filled the world with other metropolises: Tokyo, Delhi, Shanghai, Cairo, Beijing, New York City, Istanbul, and Moscow. Most of us now live in cities. But even as we gravitate toward large city centers, Scripture points us to a better city to come.[80]

And so we wait. Waiting is relevant in our age of technological marvels. Faith is alive in a person who sees the advances of mankind—the cities, the innovations, and the future tech—and still hopes in an invisible city to come. The Christian lives between the

---

78 Daniel J. Treier, "City," *Evangelical Dictionary of Theology*, ed. Daniel J. Treier and Walter A. Elwell (Grand Rapids, MI: Baker Academic, 2017), 190.

79 Jacques Ellul, *The Meaning of the City* (Eugene, OR: Wipf & Stock, 2011), 16.

80 Rev. 21:1–4.

"already" and the "not yet" of our salvation, a contented delight in our present salvation with a hopeful expectation of a coming heavenly city made by God. We will look at this new city more carefully later. But God's story of creation is his work to create a people, and he will create a delightful city so that his people can enjoy his created city forever, and so that he can delight in the joy of his people as they delight in what he has made.[81]

So right now we are faced with the question: Am I fixated on earthly technologies and earthly cities designed and engineered by the innovation of men and women? Or is my hope set on a city to come, designed and built by God himself? Babel shows us how easily we sinners set our immediate aspirations on the products of our own innovation and creativity. Abraham shows us a different way, the way of faith. And he shows us why, even in this age of innovation, we should feel low-grade discomfort. We don't feel at home in this technological age.

Takeaways

Technology represents a major part of the drama of human history. But it has limits. Soon we will look at what technology cannot accomplish. But first we should gather a few takeaways from this point in the discussion.

1. We only discover what God infused into creation.

Humans are great at dreaming and hypothesizing about fanciful possibilities of what could be, but all those ideas are confined to what is possible inside the created order. The humbling truth is that humans "cannot introduce anything into nature; they can only

---

81 Isa. 65:18–19.

derive things from nature, since it is not humankind but God who causes it to be present in nature and who created nature." Only God's intention can make scientific discovery possible. Within that history of innovation is the reality that "humans toiled for many centuries with extremely inadequate tools while all we needed for the advancement of our undertaking lay at our feet all along, as it were. But we are not aware of it; in effect, we are blind to it." Until at some point, in God's timing, he points us to a new innovation or discovery. And then "science and learning begin to boast about their discoveries and put on airs as if they were the actual ones that accomplished this, even though with all their findings they could not create or produce anything; they could work only with powers that they had discovered as they had been created by God."[82] This is the basic plotline of human innovation.

At every step in technological innovation, new breakthroughs give us a glimpse into the mind of the Creator. If human discovery is simply how we uncover the latent possibilities and inevitabilities coded into creation, we should marvel—not at human innovation but at the Creator! When the Wright brothers finally got their cloth-covered plane to skim the surface of earth for a few hundred yards, God was not sitting back enjoying a novel human innovation. And he was not watching the end result of a plane that could simply be attributed to years of trials and errors, wins and crashes. No. In that moment, he was watching humans apply something explained 165 years earlier with the Bernoulli effect, another discovery into the patterns coded into the creation intended for us to discover and use later. How did the Wright brothers fly in 1903? How can we fly today? Because the Creator taught us.

82 Kuyper, *Common Grace*, 2:586.

This is why evolutionist Kevin Kelly can say, "The technium can't make all imaginable inventions or all possible ideas. Rather, the technium is limited in many directions by the constraints of matter and energy."[83] Yes, but we must add that these constraints are God-given.

If we invent anything that betters our lives, it comes from the common grace of God, who patterned that possibility within creation from the dawn of time and, by his Spirit, created people to work toward a discovery in the Creator's own timing. When we make a tech discovery, we are simply learning what God already knew from the beginning. We discover the latent possibilities and the inevitabilities coded into creation by the Creator. What untold millions of possibilities in this globe and in space exploration remain undiscovered?

2. God sends innovation through geniuses and by inevitability.

Debates over the origin of human innovations have led to two dominant theories. Some people support a "heroic theory of invention" and say that new innovations come from the geniuses who discover them. In other words, a discovery would not be made without a unique discoverer, like an Einstein or an Edison. Other people support a theory of "multiple discovery" or "simultaneous invention" and say that at any given time in the unfolding of a tech tree, there is a moment of time for an inevitable discovery, so inevitable that multiple people will make the same new discovery, independently, remotely, at roughly the same time.[84]

The late cosmologist Stephen Hawking wasn't religious, and his concepts of God existed in an abstract and impersonal scientific

---

83 Kelly, *What Technology Wants*, 119.
84 See a fine survey of simultaneous innovations in Kelly, *What Technology Wants*, 131–55.

place. Yet at the end of his life he said, "Knowing the mind of God is knowing the laws of nature."[85] That's absolutely true, if not exhaustibly true. As we discover natural laws, and what those laws make possible or impossible, we make further discoveries into the patterns of creation put there by the Creator himself. A similar point was made by Thomas Edison, an inventor who was awarded more than a thousand patents in his illustrious career—including the incandescent light bulb, the phonograph, the alkaline battery, and the X-ray fluoroscope. Edison was a freethinker, more agnostic than believer. But he did believe in nature and once admitted: "I've got no imagination. I never dream. My so-called inventions already existed in the environment—I took them out. I've created nothing. Nobody does. There's no such thing as an idea being brain-born. Everything comes from the outside. The industrious one coaxes it from the environment."[86] So where do inventions originate, if not in the inventor's head? They're enticed from creation. They come from the LEGO drum and are often built at the same time by multiple inventors.

Isaiah 28:23–29 is a great example of simultaneous invention. The farmer was not a heroic inventor. He wasn't Einstein. He wasn't even one of Cain's famous great-great-great-grandkids. He was just a simple farmer, and God taught him to plow, seed, and harvest. The text implies that God was teaching thousands of other farmers the same skills, at the same time, all across the globe, and perhaps through more advanced or rudimentary tools. God sets the timing for when innovations are discovered.

Steam power was understood long before it was first put into operation in 1712. But there's no reason why steam power couldn't

---

85 Stephen Hawking, *Brief Answers to the Big Questions* (London: Bantam, 2018), 28.
86 Edmund Morris, *Edison* (New York: Random House, 2019), 12.

have been harnessed this way centuries earlier, writes Kuyper. Steam's long delay "shows poignantly how it is God himself who guides all of human affairs, giving human civilization an entirely new impulse only when, in his counsel, it was destined to happen."[87] The timeline for our inevitable innovations awaits God's schedule. God alone governs the volume, speed, and timing of our scientific discoveries and technological trends.

This holds true in the digital age. When asked why he created Twitter, cocreator and CEO Jack Dorsey admitted he didn't. Twitter "wasn't something we really invented," he said, "it was something we discovered." He and his team simply deployed early SMS technology, used in texting, to broadcast text messages, not to a chosen few, but to a limitless many. The small Twitter team began using the new tech to update one another's whereabouts live when they were out of the office. "It was such an amazing feeling knowing that I would be sending an update and potentially it would buzz [in] some of my friends' pockets and they would take it out and they would understand—in that moment, immediately—what I was going through, what I was thinking." The instant connection knit the team together. "It felt electric. It felt very powerful to us." But the potent digital platform didn't emerge from nowhere, ex nihilo. Twitter was discovered. In due time, Twitter was inevitable.[88]

So which theory is right: the heroic theory of invention or simultaneous invention? Scripture seems to settle the debate by showing how both operate concurrently. In Cain's lineage, we see the *heroic* with select individuals who mark the genesis of new fields of human innovation. And in the ancient farmer we see the

---

87 Kuyper, *Common Grace*, 3:498.
88 "Jack Dorsey on Twitter's Mistakes," *The Daily* podcast, August 7, 2020.

simultaneous, with disconnected farmers across the globe discovering new farming practices concurrently. Within God's timeline for human invention, both theories work in concert.

3. Technological progress moves along the Creator's guide rails of possibility.

Multiple discovery happens when humans take their cues from creation, often at the same time, but not always in the same place. The long process of human science, engineering, and technological advance unfolds as man engages his intellectual powers to read the possibilities inherent in the created order. God gives man the intellectual powers to discover creation's possibilities, and—thought by thought and experiment by experiment—science, engineering, and technology advance. New discoveries are not accidents, nor merely the product of human brilliance. God governs the history of technological progress by establishing the guide rails of possibility in creation through the scarcity and abundance of raw materials and by the limits of natural law. First, the Creator sets the boundaries of the sandbox, and then he reveals to man the tools to dig and discover and design inside it.

This is where Darwinian evolutionists become very helpful, because while I deny that their theory is a fitting way to understand biological history, I think that evolutionists give us the right language to speak of technological evolution. Technologies build through combinatorial evolution. Old technologies merge into new technologies, but they also evolve according to the "constant capturing and harnessing of natural phenomena." All human innovation helps exploit or resist natural laws. "A technology is always based on some phenomenon or truism of nature that can be exploited and used to a purpose." It's a rather obvious statement but truly remarkable in its implications when Arthur says, "Had we lived in a universe with different phenomena we would have had different technologies."[89] The natural patterns of *this creation* led to *these technologies* we now have in hand.

Thus evolutionist Kevin Kelly says, "Our role as humans, at least for the time being, is to coax technology along the paths it naturally wants to go."[90] Human innovation is like a dam of inevitability, "like water behind a wall, an incredibly strong urge pent up and waiting to be released."[91]

Creation's potential does not mean that technological discoveries are free from the sinful desires of man. They never are. Yet within creation itself comes the inspiration we need for new discoveries. From the opening chapters of Genesis, humans responded to creation by birthing new industries. Our native impulse to cultivate creation seems to work in concert with a pent-up force inside creation that calls forth new discoveries. The possibilities inside creation order our behaviors and shape our techniques. Creation holds within it something of a blueprint for human innovation and an ordered timeline for when certain innovations will emerge. We invent because the creation itself excites certain inventions from us at the right time.

Human tech is only made possible by the possibilities of creation. Call it "discovery," call it "invention," call it "innovation"—human behaviors are precoded into creation by the Creator and discovered by humans through experimentation.

God and his creation expect our innovation. In Psalm 104:14–15 we are told that the Creator caused plants to emerge from the soil in order for us to have bread, wine, and oil—because God intends to

---

89 Arthur, *Nature of Technology,* 22, 46, 172.
90 Kelly, *What Technology Wants,* 269.
91 Kelly, *What Technology Wants,* 273.

gladden, invigorate, and strengthen our bodies. God's aim began with three organic gifts (grain, grapes, and olives). Each gift corresponds to an intended outcome (bread, wine, and oil). Between gift and outcome is the human process or technique (baking, fermenting, and pressing). The Creator reaps all the praise for the effects of our bread, wine, and oil (to gladden, invigorate, and strengthen our bodies), not merely because he patterned the raw ingredients we needed, but because he raised up the exact right plants to produce these refreshing end results. The human techniques we use to produce the end results were coded by the Creator into creation itself. Natural resources and natural laws are guide rails that limit our innovative possibilities and providentially shape the trajectory of what innovations work best. So it's entirely right to say that the farmer's farming techniques are learned directly from the Creator via creation. We can say the same about the baker, the vinter, and the olive oil maker. The voice of the Creator, heard in creation, calls forth human skills.

4. Creation is an open-source playground rated MA.

God created the earth and he called it good. He shot electricity through the sky, scattered uranium deposits inside the soil, and stuffed genetic codes into cells. Thermodynamics was his idea, nuclear fusion was his design, and nitrogen supplements was his provision. All of this technology was intended to be discovered. For good or ill (and we must soon talk about the ill), the Creator handed us an open-source creation. He gifted us with the periodic table of elements, and with what seems to us to be an infinite number of ways to mix and match elements of the created order. This world is freighted—and a little frightening—with possibilities. The Creator gave us fire, electricity, coal, copper, meteoric iron, oil sludge, uranium, and the genome to investigate. If human discovery into some of these areas makes you uneasy, it should. I got the same feeling the first time someone put a rifle into my hands as a small boy on a farm to shoot at targets. The feeling of being ill-prepared for the potential power I held reminds me of humans holding the powers of new innovation. We are finite little children, given not simply a bucket of LEGOs or a loaded rifle but a pile of hand grenades. We've been entrusted with explosive powers that can destroy bodies and maim creation. We wield powers that require great diligence and wisdom.

And yet we are mostly talking about materials within reach of shovels, backhoes, and drilling platforms. I cannot imagine what's infused deeper in this globe, not to mention in the worlds around us—new energy and new metals for billions of humans to flourish on this planet, and on other planets, in the centuries ahead. The right way to make new discoveries is not with the greedy fist but with the open hand, as if receiving a gift.

5. Technology is from dust and returns to dust.

Whether it's bringing the sacrificial lamb for the offering or creating the wine and bread used for Communion, even man's most holy creative acts are fundamentally acts of destruction. We make, and therefore animals, grapes, and grain must die.[92] But most of our technologies are made from the inanimate, taken from the dirt in the raw materials of this earth—excavated, refined, and innovated by the earth's caretakers (us). Our tech, like ourselves, is taken from the ground and will return to the ground as dust. Inventions are not eternal but sprout on a technology tree in the timeline of creation. They each share a life cycle, similar to humans. God's creation of Adam and Eve was a paradigm of human innovation. Adam and Eve were taken from the

---

92 Gunton, *Christ and Creation*, 125.

ground, given life from God, and commanded to work the ground. Because of their sin, their bodies will reenter the ground.

Our gadgets come from the ground, and in a manner of speaking, we breathe life into them. Ideally, those technologies serve the flourishing of humanity before they are eventually put back into the ground or recycled into something else. But a futility cycle taints all our innovations. Ecclesiastes speaks to our tech age when we are told: "What has been is what will be, and what has been done is what will be done, and there is nothing new under the sun" (Eccles. 1:9). We cannot escape the futility. None of our technology will escape the allure of human power over everything. We are blind to our own vanity through the mirage of technological progress. We are not moving forward as much as we are circling around in a loop of nothing-new-ness that cannot redeem us. The honest reality that should govern our aspirations (along with our consumptions) is that all our innovations are ultimately headed to a recycling center or a landfill. They cannot serve as objects of our heart's enduring hope.

6. God's provision is visible at the micro level.

The life cycle of innovation doesn't necessarily mean that new technologies generate more waste. In one sense, technology makes us less reliant on certain bulk raw materials. Think of vinyl music records replaced by cassettes replaced by CDs replaced by intangible MP3s. And yet, as our technological advances shrink into digital nothingness, we lose sight of created realities and take for granted the physical resources required of the planet.

On May 31, 1988, President Ronald Reagan traveled to Russia to help ensure that the Berlin Wall would fall (and it did, sledgehammered down seventeen months later). But in this trip, Reagan argued that the wall needed to fall because all humanity was on the brink of a global "technological revolution," as he called it, a revolution of "the tiny silicon chip, no bigger than a fingerprint," a chip that could replace rooms full of old computer systems. Humanity was about to experience a techno-rebirth. "Like a chrysalis, we're emerging from the economy of the Industrial Revolution—an economy confined to and limited by the Earth's physical resources," and into a mind economy, he said. "Think of that little computer chip. Its value isn't in the sand from which it is made, but in the microscopic architecture designed into it by ingenious human minds. . . . In the new economy, human invention increasingly makes physical resources obsolete. We're breaking through the material conditions of existence to a world where man creates his own destiny."[93]

The events were not unrelated. The democratization of the computer chip in personal computers and then in smartphones would doom communism.[94] Perhaps the president knew this. But Reagan's bold words echo a false assumption about micro technologies: that as our tech becomes smaller, the natural elements themselves lose importance. I think this logic is flawed. First, shrinking technology never lessens demands on natural resources. Even as our gadgets shrink, we generate mountains of consumer electronics waste every year. Second, even as technology shrinks, we never transcend the richness of the natural resources of earth. Instead, we become more appreciative of them, especially the rarest elements. Micro innovations don't diminish creation but instead make us more dependent on earth's most precious elements. New tech creates the need for more mines, not fewer.

Think of the material glories in small proportions, like microscopic metal threads, four micrometers in diameter, which are

93 Ronald Reagan, "Remarks and a Question-and-Answer Session with the Students and Faculty at Moscow State University," reaganlibrary.gov (May 31, 1988).
94 Harari, *Homo Deus*, 377.

coated with a corrosion-resistant polymer and poked into the surface of the brain to make neuro-machine interfaces possible. Or think of the smartphone. The iPhone takes its cues of possibility from the created order. From screen to battery to processor to sound to casing—each smartphone blends together more than sixty elements.[95] In the words of one friend in the tech industry, "I find it amazing that all those obscure elements in the periodic table that I had to learn in chemistry class in high school in 1977, a task that seemed pointless at the time, are now critical to the smartphone age. And the elements were all there from the beginning! God created a vast world of diverse elements, crude oil, and silicon, even though for almost the whole of history these things were scarcely used." And if we could not have predicted the iPhone forty years ago, imagine what's in store for our future. Whatever technological wonders lie ahead, they're already latent within the periodic table, within the ground, within the sandbox that God gave us.

In the availability of micro elements, we may be unable to conceive of limits to our innovative possibilities. But the limits remain. And no matter how small our technologies shrink, we cannot lose sight of the material elements that make every new achievement possible in the first place.

7. Technology is God's gift to mankind to push back the curse.

A farmer knows more than anyone that technology is God's gift to push back the curse. Remember the curse God pronounced on Adam, which cursed agriculture and every farmer who would be born thereafter:

---

95 See Andy Brunning, "The Chemical Elements of a Smartphone," compoundchem.com (Feb. 19, 2014); and Jeff Desjardins, "The Extraordinary Raw Materials in an iPhone 6s," visualcapitalist.com (Mar. 8, 2016).

To Adam [God] said,

> "Because you have listened to the voice of your wife
>   and have eaten of the tree
> of which I commanded you,
>   'You shall not eat of it,'
> cursed is the ground because of you;
>   in pain you shall eat of it all the days of your life;
> thorns and thistles it shall bring forth for you;
>   and you shall eat the plants of the field.
> By the sweat of your face
>   you shall eat bread,
> till you return to the ground,
>   for out of it you were taken;
> for you are dust,
>   and to dust you shall return." (Gen. 3:17–19)

How do we deal with thorns and thistles? That's part of the technology to come in the storyline of agriculture. New farming techniques help us coax a crop from the cursed crust.

In the garden, food was readily available for Adam and Eve. After the fall, food would be pulled from the ground with great sweat and labor. This is where we see the grace of God in teaching the farmer in Isaiah 28. Technology is a gift for us to push back the effects of the fall. Indeed, "the impulse of virtually all human activity is born from the urge to combat sin or its effects."[96] This is the common grace of God in our human innovation. God placed this universe, and our own bodies, under a curse so that we would

96 Kuyper, *Common Grace*, 2:582.

live in hope of the resurrection.[97] By his grace, he also left us with innovative possibilities as a merciful gift to resist some of these effects, to heal some of the creation that is broken, and to give us new ways to manage the pain of life in a fallen universe.

Jesus reminded us that God "makes his sun rise on the evil and on the good, and sends rain on the just and on the unjust" (Matt. 5:45). All good things from creation are blessings from the hand of God. Can we spread this benevolence to the technologies we build from creation? If so, through all our very best technologies, God rains down blessing on the just and the unjust in the form of electricity, televisions, cars, jets, computers, Internet, air-conditioning, pain meds, espresso beans, and every international cuisine you can imagine. Our glut of blessings beautifully displays the goodness of God to sinners enduring this fallen existence.

8. Technologies remain under the futility of the curse.

Tech can help push back the curse, but again, tech remains under the curse. Techno-futility is inescapable. What does that mean for us?

Human innovation is loaded with ironies. If you dig a pit, you can fall into it.[98] If you quarry stones, they can roll down on you and crush you. If you split logs, limbs can get in the way of the axe.[99] Or as Paul Virilio noted, "When you invent the ship, you also invent the shipwreck; when you invent the plane, you also invent the plane crash; and when you invent electricity, you invent electrocution. . . . Every technology carries its own negativity, which is invented at the same time as technical progress."[100] This

---

97 Rom. 8:18–25.
98 Ps. 7:15.
99 Eccles. 10:8–10.
100 Paul Virilio, *Politics of the Very Worst* (Cambridge, MA: MIT Press, 1999), 89.

is the danger of our open-source creation. All our technological progress remains under the curse. So should we just stop creating technologies? Because then the world would be safer? No. While Virilio has a point, the reality is that sky-related deaths can be traced back before flight inventions to asteroids and volcanoes that predate the Wright brothers (think of Pompeii). Drowning by catastrophic flood predates the *Titanic* (think of the flood). And death by lightning strike predates the electrification of cities (think of wildfires). Creation's curse predates Chernobyl by millennia. Mortal dangers, in every direction, are the consequences of life in the fallen world, whether we create inventions or not.[101]

Tech pessimists will say that new innovations always introduce more problems than solutions. Tech optimists will say that the best innovations solve more problems than they create. But both agree that new innovations bring new problems and new complexities. That's part of the cursed reality of life in the fall. So while technology is a gift from God to help us deal with the daily aches and pains of life, in the end those same technologies remind us that we still live under the curse. In his sovereignty, God knew that the potentials he loaded into creation introduced both rewards and risks. That's because, in the process of inventing constructive technologies, his fallen image bearers will also discover destructive technologies. As we looked at in the last chapter, God is sovereign even over harmful technologies—though the moral consequences fall not on the Creator but on the wielder. Part of the

---

101 Also commonly true of older innovations. As Spurgeon said in an age fearful of trains, "If the steam engine had never been known, and if the railway had never been constructed, there would have been sudden deaths and terrible accidents, notwithstanding. In taking up the old records in which our ancestors wrote down their accidents and calamities, we find that the old stagecoach yielded quite as heavy a booty to death as does the swiftly-rushing train." C. H. Spurgeon, *The Metropolitan Tabernacle Pulpit Sermons*, vol. 7 (London: Passmore & Alabaster, 1861), 481.

curse we are forced to carry in this fallen, technologically advanced planet is that we cannot un-invent anything.

9. Cutting-edge advances will mostly come through God rejectors.

That advances come through God rejectors emerges directly in Genesis 4. In the "list of antediluvian heroes who founded the arts of human civilization, city-building, herding, metal-working, music," we see a link: "urbanization and nomadism, music and metalworking" are all tied to the lineage of Cain, suggesting that "all aspects of human culture are in some way tainted by Cain's sin."[102] Yes, early technology was tainted by Cain's sin. But that does not diminish the fact that God introduced those industrial advances into the story of humanity by Cain's lineage, by his own protective mercy and intent.

Too many Christians misunderstand this point. Cain's urban aspirations were not his veering off "to take the wrong road, where every step leads further from God," nor his "high-handed piracy of creation."[103] Quite the opposite. Cain's lineage was divinely preserved and chosen to jumpstart human innovation. Genetic refinement, metalworking, and music making—a triad of innovations that forever changed the course of human history and each of our lives—were the very purpose for preserving Cain's lineage. In the strong words of John Calvin:

> Let us then know, that the sons of Cain, though deprived of the Spirit of regeneration, were yet endued with gifts of no despicable kind; just as the experience of all ages teaches us how widely the rays of divine light have shone on unbelieving nations, for the benefit of

---

102 Gordon J. Wenham, *Genesis 1–15*, vol. 1, Word Biblical Commentary (Dallas: Word, 1998), 110–11.

103 Ellul, *Meaning of the City*, 5–7.

the present life; and we see, at the present time, that the excellent gifts of the Spirit are diffused through the whole human race.[104]

The principle of Cain's lineage remains true today. The Spirit pours out innovative brilliance liberally on those who do not trust or treasure the living God.

At the apex of their national glory, God's people were never recognized for any feats of science or engineering. Israel's boats "were insignificant compared to the fleets of Tyre and Sidon." The Egyptians, Assyrians, Babylonians, Persians, Greeks, and Romans all "surpassed Israel in every field of science and technological aptitude." On behalf of the world, Israel would contribute its Savior, not its latest mechanical advances. Innovations today may come from God's people, but on the whole, "the gifts and talents for general human development are dispensed in much stronger and greater measure to the children of the world than the children of the light."[105] That was true in Genesis 4, and it is still true today.

In his wisdom, God chooses to send the most powerful ideas and innovations into this world mostly through God rejectors, through agnostics and atheists and deists who hold to some appearance of religion (like Cain did in his offering). But for whatever reason, God has chosen to *not* send many technological advances through his bride, the church. Spiritual awakening and innovative prowess, both the work of God's Spirit, are rarely tied together in the economy of God's plan.

This is the highest human irony: to possess an increasing wealth of knowledge from the Creator and to use those powers of innovation to

---

104 John Calvin and John King, *Commentary on the First Book of Moses Called Genesis*, vol. 1 (Bellingham, WA: Logos Bible Software, 2010), 217–18.

105 Abraham Kuyper, *Pro Rege: Living under Christ's Kingship* (Bellingham, WA: Lexham Press), 1:168, 171.

trace out the Creator's patterns in his creation—while using the fruit of those innovations to dishonor God through pride and greed and power seeking. Many of the greatest, most inventive minds are also grade-A jerks, backstabbers, and users who burn through spouses, weary their children, and exhaust their colleagues and employees.

So if you were to tell me that many of the great discoverers and physicists and inventors and startup-gurus in Silicon Valley are a load of self-centered narcissists, out to grab fame and power and wealth and prestige, I'd say, "Yeah, that sounds right." I'm not saying that there are not wonderful exceptions; I'm saying that we should not be surprised if their hubris is the norm. We can still glorify God for the innovation that comes through extraordinarily gifted and powerful unbelievers, because the common grace of technological discovery continues to flow especially thick through non-Christians.

God's people benefit from the skills of unbelievers, and this is clear in the Old Testament. The Gentile king and architect Hiram, the king of Tyre, comes to mind—a man King David asked to design and build his own house.[106] Later, Solomon called on a different Hiram of Tyre, a half-Jew who learned metalcraft from his Gentile father, to lead the building of the temple in Jerusalem.[107] There was no shame in God's people calling on the skills of Gentiles. But today, many Christians want to dismiss technology on the basis of the godlessness they see inside the industry. If innovation is from Cain's lineage, we should not be surprised at the unbelief of the innovators. Instead, we can marvel at the divine intent in their lives.

More than ever before, Christians are needed in science and innovation. And more and more of the great discoveries in science *are* being led by Christians. And yet in these fields of science and

106 2 Sam. 5:11.
107 1 Kings 7:13–14.

innovation, the church will not dominate but will mostly react to innovation. In the tech age, the church is "busy with rearguard-holding operations more than with pioneering novel technologies," says Harari, intending a dig. Then the atheist asks: What innovation did a Christian pastor invent in the twentieth century that can rival antibiotics or the computer? And what does the Bible have to say about genetic engineering or artificial intelligence?[108] It's true that Christians are more reactive than innovative. But this is not a failure of the gospel's relevance; this was the Creator's pattern set in place since Cain. Harari is confused, as many are. Putting aside the mapping of the human genome, one of the great scientific advances of our age (and spearheaded by a Christian), the church does not legitimize herself to the degree that she can offer the world new innovations. The same Spirit that converts souls is the same Spirit that inspires computers, antibiotics, and a map of the human genome. Yet it remains conceivable to imagine a future in which science and cities follow such an aggressive technological trajectory that the church can only react to—extracting the good and rejecting the bad, until the whole system culminates into a global network that must all be finally put down by God at the end of human history. But we've gotten ahead of ourselves.

10. Science does not lead innovators to repentance.

It's logical to ask, Shouldn't human innovation, inspired by the Spirit, following along divinely ordained patterns of possibility in the created order, lead innovators back to God? The answer is one of the great paradoxes of the technological age. Many of the world's greatest inventors move away from God, not toward him.

108 Harari, *Homo Deus*, 276–78.

Because of the tragic, warping power of sin, the more ingenious the inventor, the more ingenious the idols that his mind creates. "Like water gushing forth from a large and copious spring, immense crowds of gods have issued from the human mind, every man giving himself full license, and devising some peculiar form of divinity, to meet his own views," writes Calvin. And "the higher any one was endued with genius, and the more he was polished by science and art, the more specious was the coloring which he gave to his opinions."[109] Tech geniuses are the most articulate idolaters. Their brilliance makes them reject revelation and cook up an elaborate god in their own innovative but fallen brains. The brilliance that leads to technological breakthroughs is the same brilliance, tainted by sin, that leads the minds of Silicon Valley to fabricate novel spiritual vanities in the worship of self-expression (Burning Man) or in a God-less vision of existence (simulation hypothesis).

Technological discovery happens only when we hear the voice of God in creation, but that voice is no guarantee that we will move toward the Creator's word in Scripture. More often it does the opposite, moving the innovator toward God-replacing vanities. That's a real paradox in the technological age. Men and women can be endowed with tremendous faculties for discovering new patterns in creation, and making new inventions as a result, but they can still remain blind to the glory of the Creator.

As a normal pattern, the world's greatest innovators do not worship the Creator. This may be why God decided, from the start, to bless non-Christians with greater degrees of innovative brilliance. That is their calling in this world, a calling that will not be given

---

109 John Calvin, *Institutes of the Christian Religion*, trans. Henry Beveridge (Edinburgh: Calvin Translation Society, 1845), 1:77–78.

to the people of God, by and large.[110] On the one hand, it's undeniably true that "the sons of this world are more shrewd in dealing with their own generation than the sons of light" (Luke 16:8–9). Yet Scripture says that all the greatest wisdom and technological discoveries of man are foolishness compared to the cross of Christ.[111] To the world, Christ's crucifixion is foolishness, a foolishness no billionaire innovator in Silicon Valley celebrates in public.

11. Technological advances exist for the church.

Let me go meta for a moment and ask: Why does the world exist? The simplest answer is that the world exists to demonstrate the Father's love for his Son.[112] The Son will have a bride, the church. And the Son will be begotten for her; he will come from heaven to earth, and he will redeem her. The world exists for Christ to have a delighted bride, all to the glory of God.

Out of the Father's love to the Son, God will show his love to the church in several ways. One way is through scientific discovery into the world through hardened sinners—not in love to the sinners, but in love to his church. God loves the church when he deploys human innovations that will provide jobs for Christians to support themselves and their families. The same tool that can destroy the sinner with worldly ambition can provide a Christian with a vocation. As Charnock says of the gifts of innovations, "These gifts are indeed the ruin of bad men, because of their pride, but the church's advantage is in regard of their excellency, and are often as profitable to others as dangerous to themselves. As all that

110 Kuyper, *Common Grace*, 1:337.
111 1 Cor. 1:18–2:16.
112 Robert W. Jenson, *Systematic Theology, vol. 2: The Works of God*, rev. ed. (New York: Oxford University Press, 2001), 48.

good which is in plants and animals is for the good of man, so all the gifts of natural men are for the church's good; for they are for that end as the principal, next to the glory of God."[113]

Plants exist to feed animals, and man. Animals exist to feed man. Everything in creation exists to serve a higher purpose than itself. This same dynamic works in human innovation. The innovative brilliance of natural man exists for a higher purpose beyond itself, namely for others to use in glorifying God in service to his mission on earth. This principle comes directly from Paul, who confirms that everything you find in this world—including, we can assume, all the gifts of innovation given to mankind—exists as a gift to the church. All innovation is a gift that reminds us not to boast in man's innovative brilliance. That brilliance is God-given and God-orchestrated to serve the church in her mission. No matter what else it does, every human innovation that benefits the world is a gift from God, for his glory, in service to his people.[114]

As I write, my family is locked at home in a coronavirus quarantine, looking forward to "gathering" with the people of God virtually on Sunday through YouTube live video. As I think of this temporary situation, my mind wanders to how many millionaires (and billionaires) it took to create YouTube, not to mention the number of men and women who will ultimately be ruined by the wealth and power they've been given by mastering this video platform. And yet what if God ordained all of it to create a cutting-edge and reliable video platform for the church to use redemptively? If the social-media makers rake in billions of dollars, that's a temporal consequence of the technology existing for the bride. These platforms destroy "bad men" with worldly wealth. But they serve the church and so become necessary.

113 Charnock, *Complete Works*, 1:67–68.
114 Rom. 8:28; 1 Cor. 3:21–23.

The world is filled with captivating gadgets and tools and media—all as an overflow of possibilities. And the church is called to discern which of those gadgets, tools, and media actually serve her mission as she seeks the kingdom of God. So we need discernment to know when God has allowed technological possibilities to exceed the necessities of the church. But here's the bottom line: God-rejecting innovators are given their gifts, not as God's love to the innovator, but ultimately as God's gift to the church. It's not unlike the metalworking brilliance of Cain's lineage, a divine gift that Noah used to build and hold together his ark.

12. Technological abuse must heed the voice of creation.

As we juggle all these possibilities of technology, we will make mistakes. In this fallen world, our technologies will damage creation and each other. This is part of the human condition. We are called to invent, and that includes the mistakes we must learn from. To be free to create is to be free to fail.

Not only will we make mistakes, but our planet will undergo major changes too—as our population continues to grow from seven to ten billion people. Certain animals will go extinct. Native habitats will be disrupted. All the life cycles and patterns of earth will be changed. In other words, "ecology will *always* be a problem. All human activity changes the balance of what there is, and should do so, unless we are to see the aim of activity as the achievement of static equilibrium."[115] It's not. Humans necessarily disrupt balances. That's what we do.

But here we can stop and appreciate that risk is amplified in the tech age. Faulty medications, faulty nuclear plants, faulty

---

115 Gunton, *Christ and Creation*, 125.

hydroelectric dams, faulty containment around experimental pathogens—all of these technologies are amplified risks with far greater fallout. So as we innovate, we must listen carefully to creation. It will remain hard *not to hear* creation. I hear it now as I write, locked down at home in a pandemic, quarantined by a virus in a desert city without a measurable rainfall for 140 straight days in the hottest-ever-recorded summer. I hear it as over five hundred wildfires burn millions of acres of forest in California, sending smoke into skies over the state's large cities to make them glow dark red in the noonday sun as ash rains down like snow. I hear it as a tropical storm hits Texas and Louisiana one day ahead of a category-four hurricane to hit the same land, twin storms gestated simultaneously in the Gulf of Mexico. No matter how buffered we may feel by our innovations, it will be a long time before we stop hearing the groans of creation.[116] Something in this world is broken beyond fixing or self-healing. No matter how far we innovate, we cannot escape nature's fury. Perhaps we invite her fury.

Over time we become more susceptible to nature's changes. When ash from a volcano in southern Iceland shut down air travel all over Europe, Slavoj Žižek noticed that the economic fallout existed only *because of* air travel. "A century ago, such an eruption would have passed almost unnoticed," he said. "Technological development has made us more independent from nature and, at the same time, on a different level, more dependent on nature's whims."[117] Only commercial jets can make our economies susceptible to nature's mood swings at 30,000 feet.

Nature disrupts us, and we disrupt her back. We disrupt creation *necessarily*, and we disrupt creation *unnecessarily*. And we

116 Rom. 8:22.
117 Slavoj Žižek, "Joe Public V. the Volcano," newstatesman.com (Apr. 29, 2010).

need discernment to untangle the two. We need scientific debate, dissenting voices in our dialogue with creation, so that over time we learn the pros and cons, uses and misuses, help and harm of our new technologies. As we imagine, make, test, and unleash new technologies into the world, adjustments will always be needed.

If we are willing to listen, cancers will tell us when our technologies have gone too far for the human body. Air and water pollution will tell us when our technologies have gone too far for the planet. "Humans were always far better at inventing tools than in using them wisely."[118] Initially this is often true, but not ultimately. Humans show resiliency in overcoming the abuse of tools and in correcting for carcinogens like food additives, chlorofluorocarbons, asbestos, and lead paint. If we are willing to listen, we will heed and correct and scale our ambitions to the health of our bodies and the health of our planet.

13. All innovation should fix our awe and thanks on the Creator.

We also abuse technology when we use it to push God further out of our daily lives. As our innovations reduce danger, they also reduce prayer, faith, and gratitude. The ancient farmer brought sacrifices to God. But now, modern industrial farming "has no use for gods, because modern science and technology give humans powers that far exceed those of the ancient gods."[119] Ancient sailors, in rickety vessels and without tools of navigation, marked their perilous ocean voyages first with "propitiatory sacrifices," then, with any divine luck, they returned home "ornamented with wreaths and gilt fillets to thank the gods" in the nearest temple. Steam ships changed all that. Steam ships "killed all gratitude

118 Yuval Noah Harari, *21 Lessons for the 21st Century* (New York: Random House, 2019), 7.
119 Harari, *Homo Deus*, 99.

in the hearts of sailors."[120] Safer tech offers more control, boasts greater predictability, and kills divine thankfulness.

But Christians are thankful people, or at least we should be. We can resist the temptation to forget the Giver for the lure of more and more powerful gifts, to neglect the Creator for the control of our own little worlds. We show our gratitude to the Giver by refusing to become addicts to his gifts. Instead, we pray for the wisdom to use his gifts in a spirit of Godward gratitude and restraint as the precious things he has blessed us with—like the smartphone and the potent digital access we have to one another. We are not called to find our comfort in controlling this world. Life isn't about embracing every comfort and controlling every variable. If personal comfort is the driving motivation in your adoption of technology, it's a worship-killing trap. But all the technologies that make our lives more comfortable—dishwashers, air-conditioners, safe cars, and electrified cities—these remain tremendous gifts from God, and he gets the glory for patterning creation so these gifts can be ours. So we pray: "God break me free from the idols of comfort, and fill me with God-centered awe for the gifts in this world that you have given me to use and enjoy." All of the patterns and possibilities in creation are divine gifts which we must steward carefully.

With the Spirit's help, we can press pause for a moment on our techno-autonomy. We can confess our comfort idolatry and the spiritual laziness that makes us lose sight of God's glory reflected in the innovations we use every day. Instead, we can recharge our souls with God-centered awe. The same God who planted the first pinecone-laden redwoods also taught the innovator how to pulp trees so that Scripture could be printed and held in our hands. The

120 Jules Verne, *Twenty Thousand Leagues Under the* Sea (New York: Grosset & Dunlap, 1917), 213. See Ps. 107:23–32.

same God who created the seas and buried a yellow-black sludge in the depths of the earth taught the innovator how to use that fuel to spin propellers and carry missionaries to unreached peoples across the globe. The same God who buried volcanoes in the oceans also scattered uranium in the ground to be excavated and refined into nuclear fission power to turn on the lights in millions of churches on Sunday mornings.

Electricity is a divine gift from the Creator. God himself crafts each lightning bolt.[121] Each of his thunderbolts could power New York City for a moment. The engineering challenge is to create a man-made lightning bolt that runs continuously, and we long ago cracked that secret. But the natural lightning bolt was patterned by God and given to us as the first cause of the digital age. Only because God patterned electricity do we now have electrified homes, electrified cities, and the proliferation of the digital age.

One of the promises God made to Israel about the promised land is that in Israel's boundaries his people would find everything necessary for their flourishing. The promised land would flow with milk, honey, bread, and olives, and it would also flourish with iron above ground and copper inground.[122] God loves to gift his people with lands loaded with great quantities of iron and bronze.[123] But God also knows that these blessings can threaten to replace him. So he cautioned his people: Beware, lest the comforts you create for yourself with these metal technologies cause you to forget me and my goodness to you in these gifts.[124] We abuse technology when we forget the Giver who gave us all these material blessings in the first place.

121 Ps. 135:7.
122 Deut. 8:7–9.
123 1 Chron. 22:3–4.
124 Deut. 8:10–14.

Many Christians struggle here, failing to inventory the tens of thousands of innovations God has given us to use every day. Many Christians, like non-Christians, sever the technologies that surround them from the grand metanarrative of God's generosity. But if God's glory shines in untouched creation (in the sun, moon, and mountains), it also shines in the innovations that concentrate and refine creation into new forms. Sixty of the earth's elements, compressed into our smartphones, give us a perspective of creation that no other generation has seen. None of our innovations are perfect. Every material gift in this life is tainted by the fall. Yet it seems that quite a lot of Christians are withholding their tech gratitude for some future innovation that will drop from the heavens, incorruptible by human misuse and without any possible side-effects.

If a tech violates your conscience, abstain from it. But if it doesn't, and you embrace it into your life, thank God for it. Give him your worship and your gratitude. Refuse to be a tech-agnostic, someone who uses the gifts but ignores the Giver. The technologist may be deaf to the Creator, but God's sheep hear his voice.[125] We can hear the Creator's extravagance in every technological gift we use—our cars, computers, smartphones, electrified homes, running water, appliances, books, magazines, plastics, Internet, Wikipedia, television, music, medicine, airliners, and Nike Air Jordans. It includes the 150,000 things you can buy in a Walmart and the 12 million things you can order from Amazon. Make a list of everything you have access to, thanks to innovation. Count up all your microprocessors, if you can. Every blessing is to be received with thanksgiving as a gift from our radically generous Giver.

125 John 10:27.

Would You Make the Trade?

In 2016, economist and professor Donald Boudreaux gave his students an ultimatum: continue living out their present, mediocre lives or trade them for the life of America's first billionaire in 1916, oil magnate John D. Rockefeller. His $1 billion net worth in 1916 translates into $23 billion today—more money than you could spend, all yours. Appealing?

With this much wealth you could own large homes all over the country, even your own private island. But in 1916 there were no private jets. Getting to each home would require traveling for days in a private rail car without air-conditioning. You could enjoy A/C in your homes, but nowhere else, not in banks or stores or the office or your friends' homes. Your chauffeur could drive you across cities in an early coupe or limo, but the ride would be slow and relatively uncomfortable. Roads were rough, and mechanical breakdowns plagued even the most basic commutes.

You could set your vision on the horizon and travel the world on slow ships, but it would take a week or more on the open ocean to arrive anywhere. As the country's wealthiest person, you could endure substandard living conditions to travel the world and taste international cuisine. But none of these delicacies would be available at home. Even for you, food would be mostly limited to local offerings. Restaurants of international cuisine didn't exist.

Your home would be electrified, but there wasn't a lot to do with the power. Beyond lamps and toasters, very few household appliances were available. There were no radio stations, no televisions. You'd have a nice record collection, but it would all be scratchy mono. You could build a personal theater, but there would be only a few silent movies to watch, assuming you could find copies. Your phone would be connected to a wall. You could buy the world's finest clocks and timepieces, but they were more beautiful than accurate.

Pharmacology and medical procedures were rudimentary. Antibiotics were not available, not even for you. Even a small infection could threaten your life. One in ten babies died in the first year, a statistic that would apply to the children born to your wife and daughters. Dental procedures couldn't save your teeth. Dentures would be inevitable. There were no contact lenses.

"Honestly, I wouldn't be remotely tempted to quit the 2016 me so that I could be a one-billion-dollar-richer me in 1916," says Boudreaux. In contrasting the comforts, "nearly every middle-class American today is richer than was America's richest man a mere 100 years ago."[a] Our safe jets, reliable cars, intelligent phones, medical options, household appliances, streaming video, digital music, endless comforts, and uncountable consumables have upgraded each of us to a tech wealth beyond Rockefeller's wildest imagination.

a   Don Boudreaux, "Most Ordinary Americans in 2016 Are Richer Than Was John D. Rockefeller in 1916," cafehayek.com (Feb. 20, 2016).

The Answer

So was uranium scattered in creation by accident or by intent? Did atomic power get put in our hands by chance or by purpose? Did God forget to hide the petroleum reserves deep enough to keep them from our reach? Did someone forget to childproof this place?

The technologies in our hands spring from the patterns of the earth. The Creator controls the raw materials put into the ground for us to discover and use. He controls the natural laws for the technologies we create. He gives us scientists who explore the patterns and innovators who exploit patterns into new tech for us to use. The process works because it follows the voice of the Creator.

God makes things out of nothing. We make things from what is available. He is sovereign; we are limited. Yet we've managed to invent potent innovations. We have multiplied powers at our disposal to meet many of our needs and wants. But technology cannot solve our greatest need.

4

What Can Technology Never Accomplish?

GOD CODED EVERY TECHNOLOGICAL possibility into the created order. And inside each of us he coded the desire for transcendence. God "has put eternity into man's heart," which means that we are always looking for more.[1] This world is never quite enough. So it's no surprise that the first human endeavor was to build a tower to heaven, to enter the heavens, to travel to space. We are hardwired to build tall towers and fire huge rockets to transcend this planet.

The motivation for our aspiration is all around. On a recent spring evening, my family went outside in the backyard to gawk into the clear night sky. The largest supermoon of the year was at its biggest and brightest, like a spotlight in the black sky—God's creation, shining radiantly in the dark. About fifteen minutes later, we looked in the opposite direction and watched the International Space Station zip across the sky at 17,000 mph, 250 miles above our backyard.[2]

---

1   Eccles. 3:11.
2   The evening of April 7, 2020.

The glory of God shone in the heavens to my right. The glory of man whizzed across the sky to my left.

In 1958 the Soviets put the first satellite in space. For the first time, humanity could look into the sky and lay claim to a technological device in the heavens. Sputnik sparked the hope that one day, we humans could escape the confinement of this earth, and a decade later, Neil Armstrong and Buzz Aldrin pressed their boots into the moon's dust. Apollo 11 captured the attention of the world. Space travel reached Babel-like heights, drawing together the collective aspirations of mankind. Finally, some said, man could cast off the old traditional myths of this planet and reach for new planets. Man could discard the old, dated, religious rituals and grab for immortality through scientific discovery and space travel.[3]

Moments after the Apollo 11 moon landing, an excited sci-fi novelist saw the launch of worldwide peace. Ray Bradbury thought that the rocket would reunite mankind. "Because when we move out into the mystery, when we move out into the loneliness of space, when we begin to discover we really are three billion lonely people on a small world, I think it's going to draw us much closer together," Bradbury told Mike Wallace on television. Space travel would energize mankind in ways only warfare could previously:

> We've always wanted something to yell and jump up and down about. And war is a great toy to play with. Men and boys loved war. They pretended at times that they don't love it, but they do. Now we've found a greater love, one that can bind us all together, one that can fuse the entire race into one solid mass of people

---

3   On the captivating awe of space travel in light of the drabness of religion, ultimately leading to disillusionment of finding nothing on the moon, see "Moondust," *The Crown*, season 3, episode 7, produced by Netflix, November 17, 2019.

following a single ideal. Now let's use this thing. Let's name this ideal and let us eliminate war because the proper enemy is before us. All of the universe doesn't care whether we exist or not, but we care whether we exist. Now we've named the universe as the enemy and go out to do battle with it. That's the big enemy. And this is the proper war to fight.[4]

On live TV, humanity together watched the first moonwalk. But it didn't unify nations. Nor have other cosmic discoveries since. In multiple space missions, astronauts have experienced what is called the "overview effect," the metaphysical shock of getting distance from the earth and turning back to view our spinning globe as a whole—a boundary-less unity, shielded by a thin atmosphere, and surrounded by a vast emptiness of blackness in every direction. The "overview effect" causes astronauts to question the territorial wars and international tensions of humanity. If everyone could see the earth from space, they say, we would get along. NASA may offer "extremely expensive mystical experiences to astronauts," but the buzz never translates into global peace.[5] From a distance the world looks peaceful, but it's rocked by localized human discord.

It's true that since man stepped on the moon, there have been no world wars. But this relative global peace is due to the rise of the knowledge economy, says Harari. As science and innovation became a nation's most lucrative resources, wars became more limited to the few regions where material-based economies remain. Wars today are most likely to happen near Middle East oil fields or Rwandan coltan mines. But for many countries, human knowledge

4   Tony Reinke, "Ray Bradbury on Space Travel," tonyreinke.com (January 28, 2020).
5   Wendell Berry, *Essays 1993–2017* (New York: Library of America, 2019), 140.

is its greatest treasure. And you cannot invade a nation with an army of foot soldiers to loot its knowledge.[6]

More drastic than a world war, the planet itself is dying and becoming a liability to the future of humanity. Or so we are told. And yet there's an optimism in today's space explorers like Elon Musk. When asked, "What will make the world a better place?" he replied: "To wake up in the morning and look forward to a future where we are a spacefaring civilization, out among the stars, is very exciting. Whereas if you knew we were forever confined to earth, that would be very sad."[7] Creating a new ark in the form of a SpaceX rocket to take us to Mars is not a hobby for Musk. According to his biographer, Mars is the comprehensive "sweeping goal" and "unifying principle over everything he does."[8]

Space travel draws the technologists out of bed in the morning. It sparks our creativity. It will unify humanity, end warfare, and deliver happiness and meaning to our lives. It will provide humanity with an escape pod before our oceans boil away, our boulders vaporize into nothing, and our planet is sucked into the sun. It is our only hope to save us from this failing planet. By space travel we will save ourselves.

These are the technological hopes of man. We have high trust in rockets.

But even as Neil Armstrong took his "giant leap for mankind," John Updike recounted the moment in one novel this way:

6   Yuval Noah Harari, *Homo Deus: A Brief History of Tomorrow* (New York: Random House, 2017), 14–21.
7   "Joe Rogan Experience #1169—Elon Musk," PowerfulJRE podcast, youtube.com (September 7, 2018).
8   Ashlee Vance, *Elon Musk: Tesla, SpaceX, and the Quest for a Fantastic Future* (New York: HarperCollins, 2015), 16.

The six o'clock news is all about space, all about emptiness: some bald man plays with little toys to show the docking and undocking maneuvers, and then a panel talks about the significance of this for the next five hundred years. They keep mentioning [Christopher] Columbus [but all I can] see it's the exact opposite: Columbus flew blind and hit something, these guys see exactly where they're aiming and it's a big round nothing.[9]

We're surrounded by 100 billion to two trillion galaxies in a physical space stretching across at least ninety billion light-years. Inside this unfathomable expanse we call one tiny blue marble home. We can propel ourselves at 25,000 mph to another huge rock, but there we will find just another rock—another big, round nothing, another reminder of the desolation and the emptiness. To some, these new globes are blank canvases for human self-making. But even with this longing for transcendence within us, and even in the ambition of space travel, we meet more darkness.

Apart from God, space travel launches us into a bulging emptiness that continues to expand at the speed of light. As we gaze through telescopes, we are met over and over again with an endless horizon of large, unexplored, desolate globes. The deeper we peer into the blackness of outer space, the more we meet our aloneness and isolation.

Psalm 20

Like rockets bolted to a space shuttle, human hearts clamp on to innovation to find hope for the future. We see this misplaced hope in the storyline of space exploration, and we also see it in Scripture.

9   John Updike, *Rabbit Redux* (New York: Random House Trade, 1971), 23.

To understand why humankind is so susceptible to tech trust, we turn to Psalm 20, written by King David, the master technologist. Recall his famous encounter with Goliath in 1 Samuel 17, not a showdown between a war machine and a tech-less shepherd boy, but rather the story of a beefed-up war machine versus a faith-filled, tech-wise sniper. David was the sniper, and he knew how to amplify the power of the human arm with a sling. But when David penned Psalm 20 he was a grown man, well practiced in the technique of sword and armor, a celebrated king leading his army to war.

Let me set up Psalm 20. On a normal Sunday morning at my church, when you walk in the front door, you get handed a one-page bulletin printed with the order of service and responsive readings, a liturgy for the morning. Psalm 20 is like that. Specifically, it's a bulletin for a special gathering of the temple congregation to worship and pray in the final moments before Israel's king and army are sent off to war.[10]

So the king and his soldiers are outfitted for war, bearing armor and brandishing swords. But before they leave, everyone gathers at the temple to seek God's favor. The congregation begins to sing together, over the king and army, in verses 1–5.

> [1] May the LORD answer you in the day of trouble!
>   May the name of the God of Jacob protect you!
> [2] May he send you help from the sanctuary
>   and give you support from Zion!
> [3] May he remember all your offerings
>   and regard with favor your burnt sacrifices! *Selah*

---

10 For this outline I am indebted to Derek Kidner, *Psalms 1–72: An Introduction and Commentary*, vol. 15, Tyndale Old Testament Commentaries (Downers Grove, IL: InterVarsity Press, 1973), 118.

⁴ May he grant you your heart's desire
    and fulfill all your plans!
⁵ May we shout for joy over your salvation,
    and in the name of our God set up our banners!
May the LORD fulfill all your petitions!

The congregation's song and plea for God to bring victory ends here. Next, the king—King David himself, the author of this psalm—stands up to address the temple assembly with a song of God-centered confidence. Here's what he declares to the Lord in verse 6.

⁶ Now I know that the LORD saves his anointed;
    he will answer him from his holy heaven
    with the saving might of his right hand.

Psalm 20 is the king's declaration of confidence in God's sovereign governance over war. And it is true despite the spatial differences. In the expanse of the universe, and whatever distance there is between us and God in "his holy heaven"—Jehovah is near (v. 6). The infinite God of the universe is unimpeded by physical boundaries. He's omnipresent. He's always with his people. We could colonize Mars, and he is there. He is present in heaven, he is present on earth, and he is present in every one of billions of galaxies. He is not bound by time and space. We could travel light-years into deep space, and from his holy heaven he would be as present to us as he is in his temple.[11]

And yet God is also our invincible Redeemer. He is both close to us and, also incontestable, in heaven. God is so far out of the reach of mankind's collective rebellion that he is untouched by

---

11 1 Kings 8:29; Ps. 139:7–10; Jer. 23:23–24.

Babel's tower. He exists in an unapproachable light, undimmed by technological advances in any age. And yet he is also near.

Out of this theology—out of our knowledge of the sovereign, secure, gracious, and omnipresent God—we remain confident in the face of all human war tech. David proclaims his conviction in verses 7–8.

> ⁷ Some trust in chariots and some in horses,
>> but we trust in the name of the LORD our God.
> ⁸ They collapse and fall,
>> but we rise and stand upright.

Then David stepped back from the center stage, and the whole congregation finished the prewar event in verse 9.

> ⁹ O LORD, save the king!
>> May he answer us when we call.

What a scene!

Now let's skip back to David's solo about war machines in verses 7–8. Some trust in chariots and some in horses. Horse-drawn chariots "represented the most powerful military resources available in the ancient Near Eastern practice of warfare."[12] Chariots were the ancient equivalent of a tank, the most dominant power in a war arsenal. We can recall Isaiah 54 and ask: Who made the warhorse? God did. And who made the chariot maker and the chariot rider? God did. He ordained the chariot maker and the chariot rider and the chariot horse all for their own purposes. From David's ancient

---

12 Peter C. Craigie, *Psalms 1–50*, 2nd ed., vol. 19, Word Biblical Commentary (Nashville, TN: Nelson Reference & Electronic, 2004), 187.

war chariots to the F-22 Raptor, God is sovereign over every superpower of human innovation.

In Machines We Trust

So if all this is true, if God is really *this* sovereign over every technology, then perhaps—*perhaps*—we could conclude that we should
trust in those technologies after all. Because God is behind their
origin, can we therefore "redeem" these powers to protect God's
people? That conclusion would be a huge mistake. That's thinking
like the nations. In the face of war machines, what do kings do?
They grab as many as they can. They stockpile power. And yet in
stark contrast, Israel's kings are expressly forbidden from stockpiling
war machines in the form of horses.[13] Verse 7 of Psalm 20 shows
us a contrasted confidence, the confidence of the world versus the
confidence of God's people. "*Some trust* in chariots . . . but *we trust*
in the name of the LORD."

The world wears the false confidence of wealth, power, and
military might. Society walks with the unnatural swagger of self-
sufficiency: "Nothing is impossible for us!" Not so with God's
people. Our supreme confidence is in God. Carnal self-confidence
ignores God. Trust in God alone is the death of vain self-confidence,
the collapse of contrived tech-confidence. So the psalmist, in the
face of war, urges us to cast off every confidence that hinders us
from placing our exclusive trust in God.

If it seems odd to pit military power against religious faith—
some trust in ballistic missiles, and some trust in Christ—this
contradiction is not strange to the psalmist. The collective war
machines of any nation, its bombs and bullets and missiles, quickly

13 See Deut. 17:16.

become the people's hope and safety, especially in wartime. This religious-grade urgency is felt inside the global rivalry to shrink computer chips. Only smaller and faster chips can run powerful AI, a principal national security concern in the rivalry between China and the United States. AI can unleash thousands of simultaneous attacks on an enemy, far beyond the scope and speed of human response. In turn a nation's self-defense must become even more superhuman. AI will become most powerful in the hands of the first nation to corner the market of the world's fastest chips.

Whenever insecurities rattle a nation, its leaders grab for techno-security. They turn to a savior. When life is uncertain, God's word reminds us to trust in God's security, not man's armory. "The horse is made ready for the day of battle, but the victory belongs to the LORD" (Prov. 21:31). Those who look for victory in war horses and war chariots "do not look to the Holy One of Israel or consult the LORD!" (Isa. 31:1). This is man's great fail. God is the one who can give victory and the one who can bring disaster. Even with their loaded arsenals, all human superpowers are fallible flesh. God is eternal Spirit.[14] War machines are false saviors.[15]

And yet despite these cautions, "Americans worship technology," admitted sci-fi novelist Max Brooks. "It's an inherent trait in the national zeitgeist. Whether we realize it or not, even the most indefatigable Luddite can't deny our country's technoprowess. We split the atom, we reached the moon, we've filled every household and business with more gadgets and gizmos than early sci-fi writers could have ever dreamed of."[16] Yes, and when America lagged

14 Isa. 31:1–3.
15 Ps. 33:17.
16 Max Brooks, *World War Z: An Oral History of the Zombie War* (New York: Broadway Paperbacks, 2006), 166.

behind the world in telecommunications, it was only a matter of time before it took the global lead with the launch of the iPhone.[17] So when Kevin Kelly searched for Amish-like communities around the globe, he came up empty. He couldn't find another ongoing, large-scale, intentional, tech-minimalist society outside North America. Why not? Because "outside technological America the idea seems crazy."[18] Every society grasps after false saviors in the form of horses, chariots, towers, robots, rockets, or drones. But America takes tech worship to new heights.

The Gospel of Technology

Humanity may become secular, but humanity is never un-religious. Psalm 20 reminds us that human technology is more than the sum of its mechanical powers. War technology is always about faith— faith *in* something. Technology seeks to fill a spiritual void in us, to make us feel secure. In all its hopes and grand aspirations, modern technology echoes the idolatry of Babel, amplified a million times into what I call the "Gospel of Technology."

The Gospel of Technology, like the gospel of Jesus Christ, operates by its own worldview and has its own understanding of creation, fall, redemption, faith, ethics, eschatology—its own telos and endgame. The full scope of our technological saviors would require a whole book. So here is just a sketch.

The Gospel of Technology begins with man's origin in evolution. Man came from nothing, and he is accountable to no one. He has been self-improving for billions of years, and he will continue to self-create for millions of years into the future. Over the course of human history, nature has become one of our chief enemies, a

17 Vance, *Elon Musk*, 350–51.
18 Kevin Kelly, *What Technology Wants* (New York: Penguin, 2011), 231.

force out to kill our species with disasters and disease. We resist nature's killing impulse by controlling nature itself. We seek to subdue and control even the empty universe as a new place of self-preservation.

This godless, soulless, evolutionary worldview of chance and probability corrals human hope into powers of science, technique, and machines. People no longer look for perfection in the divine but in the technium, placing all progress in the hands of the technologists. Even at the earliest world fairs, when inventors boasted of the newfound powers of electricity, that electrical technology was presented in exhibitions to contrast new lighting with primitive fire making and cave dwelling—to set man's new electrified advances within a context of Darwinian advance.[19] Tech progress is simply the next step in man's evolution.

According to the Gospel of Technology, there is no fall of man, only impediments to the rise of man. The struggle is against control over myself, my image, my body, my gender, my living space, my sex expression, my life span, my productivity, my potential. Whatever hinders self-crafting must be put down. Ultimately, whatever intrudes upon each person's autonomy is the enemy, and the opposition can be defeated through innovation.

The hopes of humanity extend even to the manipulation of genetics for their offspring, in a race for the survival of the most technologically adapted. The hopes of humanity push against the boundaries of life. In a race for the survival of the most tech adapted, parents will soon begin to refine offspring with advanced genetics. And anti-aging science is rapidly developing in the field of senolytics, discovering ways to extend life by flushing the aging

19 David E. Nye, *Electrifying America: Social Meanings of a New Technology, 1880–1940* (Cambridge, MA: MIT Press, 1990), 35–36.

body of its "zombie" cells to keep the body young.[20] Many medical advances will help us care for our bodies, and we can celebrate them. But to the extent that wellness and nutrition and training become a form of self-salvation, a replacement for God, wellness becomes an idol in a false gospel.[21]

So the technologist creates a redemption of sorts. Medical technology becomes soteriology, and the body becomes a temple to the gospel of wellness. Do you want only healthy and gifted children? Genetics can help you predict and abort weaknesses. Do you seek a new gender? Modern technology makes it possible to de-gender or re-gender the body, at least at the external level of outward appearance. In many ways the tech age erodes important biological distinctions between men and women, relativizes gender, makes sex differences malleable, and ultimately washes away the value of the nuclear family.[22]

Medicine traditionally offered us the curative, palliative, and preventative. Now it offers us the augmentative. Medical technology and digital technology are merging. Brain-machine interfaces are no longer the stuff of pulp fiction or sci-fi flicks. Like Neo plugged into the Matrix, human neurological activity can now be plugged into digital receivers, human intention translated into real-time digital signals. Elon Musk's company Neuralink does this by inserting tiny flexible metal threads into the surface of the brain. More commonly, noninvasive neural interfaces will be worn as mind-reading wristbands. Thanks to these new telekinetic powers, we may one

20 Amy Fleming, "The Science of Senolytics: How a New Pill Could Spell the End of Ageing," theguardian.com (Sept. 2, 2019).

21 Kevin J. Vanhoozer, *Hearers and Doers: A Pastor's Guide to Making Disciples through Scripture and Doctrine* (Bellingham, WA: Lexham Press, 2019), 20.

22 See the story in Carl Trueman, *The Rise and Triumph of the Modern Self* (Wheaton, IL: Crossway, 2020), 225–64.

day forget keyboards, mice, and touch screens. We will control our digital worlds through verbal commands or silent intent.

If predictions are right, one day our neurological matter will be duplicated into a mindclone—either a digital scan of the brain's storage, or an augmented digital brain that watches and listens to everything we say and do in order to create an unforgettable digital database of memories (and perhaps even a digitized consciousness that will become our eternal selves).

All of this progress comes with a built-in ethic. Tech culture is acceleration culture, where optimization becomes its own end. Everything becomes about proficiency and power. Biohacking is meant to make us physically stronger and cognitively faster. It's the human toll of the acceleration economy. On the production side of our economy, technologies are called on so that we can make more stuff, cheaper and faster. Consumerism drives economic growth, and acceleration culture reinforces itself. No one asks why. The question is ignored as long as speed and efficiency keep chugging ahead. Or to say it another way, here "the commandment of love is replaced by the commandment of effectiveness and efficiency."[23] The Gospel of Technology feeds acceleration culture, but it is ultimately an aimless acceleration to nowhere.[24]

The Gospel of Technology also preaches comfort. Do whatever it takes, adopt whatever is necessary to preserve your own security and comfort in this world. Make yourself as comfortable as possible. Minimize risk. Insulate yourself from what you don't know and what you can't control. Self-preserve at any cost.

23 Egbert Schuurman, *Technology and the Future: A Philosophical Challenge* (Grand Rapids, MI: Paideia Press, 2009), n.p.

24 Hartmut Rosa, *Social Acceleration: A New Theory of Modernity* (New York: Columbia University Press, 2015).

In the end, the Gospel of Technology is the survival of the fittest. It has winners and losers, the users and the used, the adept and the naïve, the programmers and the programmed.[25] And while equality may be an ideal, inequality is inevitable.

Because technology taps into transcendent hopes and aspirations, a few endgames are beginning to emerge. In theological terms, the Gospel of Technology has an eschatology, an ultimate aim. And it branches off into two distinct forms, the embodied and the disembodied.

First, one form of technology's aim is a fully augmented embodiment. For example, CRISPR biohacking promotes the final goal of evolution, of humanity's search for a superior, posthuman species. If we can imagine a world of dinosaurs, and our intellectual superiority to them, think of a place where current humanity will be replaced by a superhuman species, and unmodified humanity (what we consider "normal" life now) will be preserved only in museums. Posthumanism will flaunt a self-made species of super-humans, a new race of neo-Nephilim, so marvelous that they will outdo primitive humanity in every way, so excellent that they will become a new race of being. Genetic technology is already climbing this transhuman trajectory. Man can now edit eggs or sperm or embryos in germline engineering, an attempt to edit DNA changes into a lineage. He will aim to regenerate through genetically fabricated "designer" babies and eventually by ectogenetically womb-less children all purified from disease, defect, and defilement.

But a second form of this eschatology is disembodied. Transhumanism is the promise that one day man will finally and fully evolve to the point of finding a way to evade this biological

25 Douglas Rushkoff, *Program or Be Programmed: Ten Commands for a Digital Age* (New York: Soft Skull, 2011).

existence, not only with a neurological plug into the digital world but with a complete transfer of human consciousness out of the brain and into a digitized awareness. Technology promises to one day exhume our brainpower out of this dying mess of biological cells we call a body.

Embodied posthumanism or disembodied transhumanism are the aspiration of the tech futurists. Tech society reaches for self-transfiguration, a manufactured new creation, and an eschatology of its own making.[26]

Evolution has become self-evolution. By the middle of the last century, with all the innovation man mustered in the Industrial Revolution, it became clear that evolution was no longer a subtle, slow-moving, invisible force guiding man over billions of years. Man had grabbed the reigns of his self-transfiguration. By 1969 Victor Ferkiss called man a "technological animal," the first creature to make technological change "the fundamental factor in human evolution." Evolution is self-directed. "Only man has evolved culturally to the point where he consciously can alter radically his physical environment and his own biological make-up."[27] Forty years later, Kevin Kelly called technology "the most powerful force that has been unleashed on this planet, and in such a degree, that I think it's become who we are. In fact, our humanity and everything that we think about ourselves, is something we've invented. We've invented ourselves."[28] Evolution has brought us here: to the pinnacle of human self-evolution, billions

---

26 "Authentically Christian eschatology needs to be distinguished from what we might call futurism, assigning roles and predicting outcomes in what is no more than a kind of eschatological technology." John Webster, *Word and Church: Essays in Christian Dogmatics* (New York: T&T Clark, 2001), 274.

27 Victor C. Ferkiss, *Technological Man: The Myth and the Reality* (New York: Braziller, 1969), 27.

28 Kevin Kelly, "Technology's Epic Story," ted.com (November 2009).

of years in the making—the technological man. He will never stop reinventing himself, remaking himself into something so glorious and superior that we can only call him posthuman.

The Gospel vs. the Gospel of Technology

If all the aspirations of the Gospel of Technology sound new, they are not. In July of 1945 C. S. Lewis published a sci-fi dystopian novel to explore this very transhumanist impulse in man. The title, *That Hideous Strength*, is a nod to Babel's ambition. The novel's transhumanist, Dr. Filostrato, seeks a world where death is evaded and where the human consciousness is exhumed from all biological limits. He is at war against the mess of biological life. The doctor calls for "the conquest of death: or for the conquest of organic life, if you prefer," in the words of Dr. Filostrato. "They are the same thing. It is to bring out of that cocoon of organic life which sheltered the babyhood of mind the New Man, the man who will not die, the artificial man, free from Nature."[29]

A month after Lewis's novel dropped, the US dropped an atomic bomb on Hiroshima, and man was reminded that eternal life is a far harder achievement than mass death. We have mastered the power to massacre. A few days after the bombings, George Orwell reflected on the novel. "Indeed, at a moment when a single atomic bomb—of a type already pronounced 'obsolete'—has just blown probably three hundred thousand people to fragments, it sounds all too topical. Plenty of people in our age do entertain the monstrous dreams of power that Mr. Lewis attributes to his characters, and we are within sight of the time when such dreams will be realisable."[30]

29 C. S. Lewis, *That Hideous Strength*, vol. 3, Space Trilogy (New York: Scribner, 2003), 173–74.
30 George Orwell, *I Belong to the Left: 1945, The Complete Orwell* (London: Secker & Warburg, 2001), 250–51.

Simultaneously, science unleashes new powers to destroy life, and new dreams of becoming eternal beings.

But for all its growing power, the Gospel of Technology stands at a face-off with the gospel of Jesus Christ. For technology to reach its final eschatological hopes, it must shove Christ aside. And now it can, says futurist historian Yuval Noah Harari. "We don't need to wait for the Second Coming in order to overcome death. A couple of geeks in a lab can do it. If traditionally death was the specialty of priests and theologians, now the engineers are taking over."[31] One gospel replaces the other by overcoming death. Technology replaces Christianity as a matter of necessity, says transhumanist Ray Kurzweil, because "a primary role of traditional religion is deathist rationalization—that is, rationalizing the tragedy of death as a good thing."[32]

First, that's a bad interpretation of Christianity. We never rationalize death. Death is our nemesis, our tyranny, and our enemy to be finally defeated.[33] We believe that one day death will be spoken about like the myth of Thanatos today, a tragedy found only in old stories like *Othello*, some unimaginable catastrophe from a bygone age.

But death is a principal battle line in the age of technology. Do we have eternal hope *through* the grave or by *avoiding* the grave? These are opposing gospels. In 1939 Lewis said that "a 'scientific' hope of defeating death is a real rival to Christianity."[34] And yet since then, "in no case has science conquered the inevitability of death. The modern man has only found more fascinating diversions

31 Harari, *Homo Deus*, 23.
32 Ray Kurzweil, *The Singularity Is Near: When Humans Transcend Biology* (New York: Penguin, 2006), 372.
33 1 Cor. 15:26.
34 C. S. Lewis, *The Collected Letters of C. S. Lewis*, 3 vols., ed. Walter Hooper (London: HarperCollins, 2004–2007), 2:262.

from sober contemplation of this fact."[35] Technological wizardry distracts man from his mortality, but it has yet to end death.

Here is where the gospel of Jesus Christ and the Gospel of Technology meet an impasse. Technologists say that we don't need someone who has died to defeat death. All we need are a couple of geeks in a lab to stop the death clock inside our cells. The savior of humanity will be humanity. We need no gods. We need engineers. We stand on the brink of self-redemption. With that belief, we come full circle and find ourselves standing again on Babel's peak among a throng of the world's strongest minds: the self-made supermen ready to storm heaven and usurp the needless God. By snapping the reigns of evolutionary development, man is in full gallop to realize Nietzsche's dream of a world where "human beings are called to transcend themselves, to make their lives into works of art, to take the place of God as selfcreators and the inventors, not the discoverers, of meaning."[36] We will all become Übermensch. Supermen. Superior beings. Self-invented. Self-glorified. Self-enthroned. Self-immortalized citizens of Babel.

For all of this technological progress, what is required? Faith. The Gospel of Technology asks only that we put our trust in the full technical control of all reality. Trust in technological improvements. Go along with them. Adopt. Adapt. Don't resist. Don't question. And don't fret if technological progress outpaces ethics. Give the technologists your faith, or at least your blind trust.

Deus Ex Machina

Technology becomes a religion rooted in an ever-deepening desire to control everything about our world and ourselves. "Just as there

---

35 Carl F. H. Henry, *Christian Personal Ethics* (Grand Rapids, MI: Baker, 1977), 47.
36 Trueman, *Rise and Triumph of the Modern Self*, 41–42.

is religious fundamentalism, there is a technical fundamentalism," writes Paul Virilio. "Modern man, who killed the Judeo-Christian God, the one of transcendence, invented a god machine, a *deus ex machina*."[37] A "god from the machine," a technological savior. Our machines will arise at the right moment to redeem us, or so we are promised.

This techno-gospel of self-evolution calls for an end to salvation in Jesus Christ. The doctrine of sin and depravity must go. Thirty years before Lewis, Herman Bavinck smelled this same cultural trend in the air. The idea "that man is radically corrupt, must be saved by Christ, and can never become holy and happy by his own power, is the most demoralizing of all the articles of the Christian faith, and ought to be opposed and eradicated with determined strength." So in gospel defiance tech-man moves toward Übermenschism, the self-redeeming reaching "not only forward, but also upward, to meet the light, the life, the spirit."[38] For more than a century, the gospel of the tech age has been the gospel of a posthuman self-deliverance.

More recently, Schuurman predicted: "The society of the future is a technical society; the ethic of the future is a system ethic; the religion of the future is the expectation of technical redemption. Humanity will trust, marvel at, and worship technology, but not infrequently also fear technical means as gods." The work of creation (*bara*) "is no longer the work of God but the work of man." Man is the only relevant maker now. The fall into sin is "not an act of becoming untrue to God but an act of becoming untrue to ourselves as humans." Redemption is not "the confession that Christ restores communion with God, but the summons to hu-

---

37 Paul Virilio, quoted in James Der Derian, "Speed Pollution," wired.com (May 1, 1996).
38 Herman Bavinck, *The Philosophy of Revelation* (New York: Longmans, Green, 1909), 274.

mans to again stand on their own two feet." Faith is replaced with "self-confidence." Freedom is not freedom in Christ, but "absolute independence." Eschatology is not about receiving a gift from God but bending the known world to our human will.[39] The world of technology breeds a culture that "recognizes no meaning or normative direction from without."[40] Technicism, our attempt to control the uncontrollable in nature and providence, leads to an unavoidable consequence, an inability to stand in awe at the God of the universe, the maker of creation. Mystery perishes, and so do relationships. "Love dies; empathy and sympathy and contact with the other disappear. Estrangement and loneliness increase."[41] The Gospel of Technology is a Faustian bargain for power, dominance, and superiority. It steals our joy, our faith, and our very lives.

Historic Letdown

Humanity's techno-utopian dreams eventually curdle into tech-dystopian nightmares. Television shows and novels abound to show how the fallout of our technology increasingly haunts the modern imagination. But predictions about how technology will let us down in the future are based on observations that technology has already let us down in the past.

Take the nineteenth century, an era that marked the most radical century of human innovation the world has ever seen. Inventions born in the Industrial Age fill a long list of gifts that we now take for granted: typewriters, cameras, film, video, conductive electricity, incandescent light bulbs, batteries, telegraphs, telephones, coffee pots, sewing machines, escalators, elevators, bubble gum,

39 Egbert Schuurman, *Faith and Hope in Technology* (Carlisle, UK: Piquant, 2003), 53–54.
40 Schuurman, *Faith and Hope in Technology*, 87.
41 Schuurman, *Faith and Hope in Technology*, 101.

Coca-Cola, high-pressure steam engines, trains, iron ships, gas-powered internal combustion engines, and automobiles. How's that for one century? And these were more than small improvements. I mean, one day you've never made a phone call, and the next day you can talk to your sister, in real time, from across town. You've never seen a picture of yourself, and then you check your hair in a fuzzy photograph. You've never used electricity or seen a light bulb, and then you walk into Paris at night and the whole city is illuminated.

As Smil points out, there were major tech advances in earlier centuries, like during the Han dynasty in China. But technological discovery accelerated in the eighteenth and nineteenth centuries due to a matrix of new factors. Inventions went global and were discovered, proposed, adopted, and diffused throughout Western culture more rapidly than ever in human history. Why? For the first time, widespread innovation was driven by projections devised in science, math, and physics. Based on past learning and confirmation, humans could imagine new physical possibilities, even before those possibilities were authenticated in labs. For the first time ever, global technological advances were rooted in scientific theories that could be consistently tested, quickly improved, and exponentially repeated. The most revolutionary part of the nineteenth century was that humans could collectively share, try, and improve their science-based discoveries.[42]

Smil offers one example:

[Thomas] Edison's light bulb was not a product (as some caricatures of Edison's accomplishments would have it) of intuitive

42 Václav Smil, *Creating the Twentieth Century: Technical Innovations of 1867–1914 and Their Lasting Impact* (New York: Oxford University Press, 2005), 7–13.

tinkering by an untutored inventor. Incandescent electric lights could not have been designed and produced without combining deep familiarity with the state-of-the-art research in the field, mathematical and physical insights, a punishing research program supported by generous funding by industrialists, a determined sales pitch to potential users, rapid commercialization of patentable techniques, and continuous adoption of the latest research advances.[43]

Especially in the 1800s, science and physics proposed new possibilities. Inventors brought those theories into reality, and industries scaled those realities and continued to improve them. The nineteenth century brought a radical diffusion of technology that had never been seen in human history. Rather than trinkets handmade by tinkerers or potions hand-mixed by alchemists, a new era emerged of chemical engineers and atomic physicists, of germ theory and thermodynamics, and of mathematicians and scientists who sought to better understand and capture and manipulate the mechanical possibilities of this world. Technology became a dynamic community project, leading Smil to name the era between 1867 and the start of World War I as "the greatest technical watershed in human history."[44]

Another way to classify this time period is to overlay it with the principles of Isaiah 28. The nineteenth-century windfall in scientific discovery was the consequence of many people finding new ways to trace out new patterns in creation. With objective theories, scientific papers, and simultaneous experiments, scientific minds around the world collaborated to better understand God's

---

43 Smil, *Creating the Twentieth Century*, 18.
44 Smil, *Creating the Twentieth Century*, 13.

designs in his creation. This was no accident. This was God's intent for science all along: patterns of creation discovered, codified, and communicated within a chorus of many men and women, in successive generations, all hearing more clearly the instruction of the Creator.[45]

No surprise, the nineteenth century also marked breakthroughs in medicine. The standard medical practices of 1800 were entirely dissimilar to those of 1899. In 1800 sickness was thought to be caused by an imbalance *inside the body* and its fluids. But the practice was so inexact that the same root imbalance could manifest in two different patients as completely different diseases. By 1899, however, sicknesses were tied to particular microbes, enemy invaders from *outside the body*. In less than a century, solutions for disease transitioned from imbalance etiology (calling for bloodletting) to invasion etiology (calling for antibiotics). The century would see the wide adoption of germ theory, leading to vaccines for malaria, cholera, anthrax, smallpox, rabies, tetanus, and diphtheria. The same century saw the first development of analgesics (aspirin), antiseptics (Purell), and anesthetics (morphine). In the train of these discoveries, chlorine bleach, long used for aesthetic purposes in the textile industry, became a prime disinfectant.

In the nineteenth century, everything changed: domestic life, health, industry, farming, communications, travel, and shipping. It was the greatest century of technological breakthrough ever witnessed in world history. The Creator's voice had never been heard more clearly in creation. So in 1900, when science journalist Edward Byrn published a five-hundred-page summary of innovations in the nineteenth century, he opened with worship:

---

45 Abraham Kuyper, *Abraham Kuyper: A Centennial Reader*, ed. James D. Bratt (Grand Rapids, MI: Eerdmans, 1998), 445.

The philosophical mind is ever accustomed to regard all stages of growth as proceeding by slow and uniform processes of evolution, but in the field of invention the Nineteenth Century has been unique. It has been something more than a merely normal growth or natural development. It has been a gigantic tidal wave of human ingenuity and resource, so stupendous in its magnitude, so complex in its diversity, so profound in its thought, so fruitful in its wealth, so beneficent in its results, that the mind is strained and embarrassed in its effort to expand to a full appreciation of it. Indeed, the period seems a grand climax of discovery, rather than an increment of growth. It has been a splendid, brilliant campaign of brains and energy, rising to the highest achievement amid the most fertile resources, and conducted by the strongest and best equipment of modern thought and modern strength.[46]

A few years later, in 1908, fifty-four-year-old theologian Herman Bavinck looked back on this same century of discovery, much of which he watched unfold with his own eyes, and wrote: "There are still many who are enthusiastic about science, and anticipate from its technical applications the salvation of humanity."[47] Yes, but did the Gospel of Technology deliver in Bavinck's day? Here's his update: "At the end of the nineteenth century, the intellectual life of people underwent a remarkable change. Although an array of brilliant results had been achieved in the natural sciences, in culture, and in technology, the human heart had been left unsatisfied."[48]

46 Edward W. Byrn, *The Progress of Invention in the Nineteenth Century* (New York: Munn, 1900), 3.
47 Bavinck, *Philosophy of Revelation*, 300.
48 Herman Bavinck, *Reformed Dogmatics: God and Creation*, vol. 2 (Grand Rapids, MI: Baker Academic, 2004), 515.

Technical creations, big houses, fruitful gardens, and the accumulation of labor-saving powers to light and water and flush and clean them—all of them combined cannot satisfy the human soul.[49] There is divine beauty to be enjoyed in creation, and there is divine creativity to be celebrated in technology. We can see God's eternal power and divine brilliance as we discover breakthroughs in science and unearth mysteries in creation. But even as they see the evidence of God, sinners grow deaf to God as their all-satisfying treasure. Even as they follow his patterns and discover his gifts, they reject him and refuse to honor him as the Creator.[50]

This is how the Gospel of Technology works. It says, "Come find safety and security in what is not God, in what we have made and invented." The distortion is older than chariots, older than steam engines, and far older than social media and smartphones with a bitten apple on them (an icon to remind us of the false promises in Eden). These same old false promises, preached in the Gospel of Technology, can neither save our bodies nor satisfy our souls. Human innovation satisfies human comforts but starves human hearts. Sinners are always trying to manufacture a new God-replacement.[51] And the latest, greatest God-replacement never delivers. If you are trying to find joy in technology—and I don't care if that's a new car, a social media platform, a gaming device, or a sexbot—the Gospel of Technology will drain your soul like a broken cistern.

No amount of innovation satisfies the heart. In fact, all the technological advances in the 1800s could not prevent two major

49 Eccles. 2:4–11.
50 Rom. 1:18–32.
51 Jer. 2:13.

world wars in the 1900s. I wonder if Bavinck saw these wars ahead? History shows us that scientific discovery does not bring world peace; it makes our weapons more powerful.

The Control Illusion

Technology promises to give us more control, and more control promises to give us more happiness. But the desire to control our lives is an illusory promise. We will never be in control. We will never become gods over anything, not even over ourselves. Big tech companies can promise us more control, but before long they will lose our private data, steal our feelings of security, and leave us feeling as vulnerable as ever.

In the tech age we can stockpile every scientific innovation that promises us more power and autonomy, but first we should pause and consider Jesus's question: For what does it profit a man to gain the whole world and forfeit his soul?[52] That is the concern, especially as the world's innovations multiply. As Bavinck says:

> In a word, agriculture, industry, commerce, science, art, the family, society, the state, etc.—the whole of culture—may be of great value in itself, but whenever it is thrown into the balance against the kingdom of heaven, it loses all its significance. The gaining of the whole world avails a man nothing if he loses his own soul; there is nothing in creation which he can give in exchange for his soul.[53]

If God is the center of your life, technology is a great gift. If technology is your savior, you're lost. The Gospel of Technology

52 Matt. 16:26; Mark 8:36; Luke 9:25.
53 Bavinck, *Philosophy of Revelation*, 257.

promises to simplify our lives and give us more free time, stronger relationships, added security, and better societies. Too often, what are we left with? More complex lives, less free time, increased loneliness, added insecurities, and amplified social inequality.

Humans are makers and worshipers, and too often what we craft with our hands becomes what we worship with our hearts.[54] Man has always been quick to bow before the idol of technology, and he always finds himself unsatisfied. Christ knew this. As the Creator of all human possibilities, he was not scientifically naïve. He authored all science! He "knew all the secrets of nature, the usefulness of human arts to the comfort of the world, but never recommended any of them as sufficient to happiness."[55] Christ created all human potential, and yet when he came to earth, he basically ignored the whole realm of technological advance in order to teach us where to find true happiness.

Jesus Christ didn't arrive as a scientist, astronomer, or inventor. He didn't incarnate to wow us with new gadgets. He didn't come to establish science; he came to be worshiped by scientists. He came to give us a greater gift. He came to give us himself. He came as our Savior.

No Control over Death

Christians are realists. We don't control death. We don't rationalize death. We stare death in the eyes and see that its tyrannical reign over us is the origin of all our sorrows and anxieties. Our chronic enemy is not the intercontinental missiles of sabre-rattling North Korea, nor is it a silent global pandemic. The last enemy that haunts

54 Isa. 44:9–20.
55 Stephen Charnock, *The Complete Works of Stephen Charnock* (Edinburgh: James Nichol, 1864–1866), 4:68.

each of us is the grave.[56] Death is the rival that attempts to torment each of us to the end. Death ends a scientist's discoveries.[57] And the fear of death tempts us to put more and more of our redemptive hopes in the hands of the technologists. But Silicon Valley will not stop the death clock ticking inside us. Anti-aging doctors may help us slow death, but they cannot postpone it. Transhumanists cannot dodge death. Only God can end death, and he has done so in and through death—the death and resurrection of his Son, Jesus Christ.

Death is powerful over us because our fundamental spiritual problem is that we fell into death's clutches by our sin. Adam's sin brought death to our entire race.[58] But we are not innocent. Every day we break the first and greatest commandment: we refuse to love and treasure God above all else in this world.[59] The essence of sin is to consider God, the maker of all things, drab and boring in comparison to the pleasures promised in the world's wealth and power and sex and technology and food and fashion and consumerism. Physical death is the well-deserved consequence for our godless loves.[60]

God broke into history to stop our suicidal addiction to what is not God. Jesus Christ—God incarnate, fully God and fully man—came to expose our fundamental problem. We refuse to be happy in the only one who can make us truly happy. We refuse to treasure God above all else.[61] Trading the glory of God for the glory of man-made gadgets is cosmic insurrection.[62] As the source of all life and happiness, God would be cruel not to command our

56 1 Cor. 15:26.
57 Ps. 88:12.
58 Gen. 2:15–17; Rom. 5:12.
59 Deut. 6:5.
60 Rom. 6:23.
61 Matt. 22:37–38.
62 Rom. 1:21–23.

happiness in himself. So he does. God "threatens terrible things if we will not be happy."[63] Specifically, God threatens terrible things if we will not be happy in him.[64] And yet we are dead in our sin. Our hearts are dead to God.

So Jesus entered his creation to redeem us. He was born under the law.[65] And by his perfect life, he qualified himself to die for us. Christ loved perfectly. He loved his neighbor perfectly. Christ loved God perfectly. We never could.[66] By his perfect love, Jesus perfectly fulfilled the law, and yet he was murdered as a lawbreaker.[67] Christ became sin for us, though he himself never sinned.[68] He took our judgment so we could be declared guilt-free for our sinful love of this world over God.

God came to us. Christ entered his creation to live a perfect life, to die a sinner's death, to be crucified, put in the grave, and resurrected to new life. "Since therefore the children share in flesh and blood, he himself likewise partook of the same things, that through death he might destroy the one who has the power of death, that is, the devil, and deliver all those who through fear of death were subject to lifelong slavery" (Heb. 2:14–15). He was the perfect, spotless sacrifice, a firstborn son, foreshadowed a thousand years earlier.[69] "Therefore, as one trespass led to condemnation for all men [Adam's sin], so one act of righteousness leads to justification and life for all men [Christ's death and resurrection]" (Rom. 5:18).

---

63 Jeremy Taylor quoted in C. S. Lewis, "Preface," in *George MacDonald: An Anthology: 365 Readings* (New York: HarperOne, 2001), xxxv.

64 Deut. 28:47–48.

65 Gal. 4:4–5.

66 Mark Jones, "The Greatest Commandment," *Tabletalk*, May 2013 (Sanford, FL: Ligonier Ministries, 2013), 14–16.

67 Gal. 3:10–14.

68 2 Cor. 5:21.

69 Gen. 22:1–19.

Through his death, Christ "abolished death and brought life and immortality to light through the gospel" (2 Tim. 1:10). Christ was raised to new life by himself, by God the Spirit, and by God the Father. The defeat of death was a truly Trinitarian work.[70]

The resurrection of Jesus Christ is the boldest historic claim of any religion. It is the essence of the gospel. The entire validity of Christianity rides on it.[71] As we believe in Christ, we have assurance in the face of our own physical death, a confidence that God will also raise us up to new life.[72] God regenerates his children and spiritually raises our affections so that we can now see and treasure the worth of God, and he gives us hope as we await our physical resurrection to come.[73] Jesus Christ broke the chains of death over our lives and welcomed us as citizens of the new creation. The Spirit fills us with joy on this earth and reminds us that we are now free from the tyranny of sin and death.[74]

Death is not simply a "technical glitch" awaiting a "technical solution."[75] Death is our enemy. We don't rationalize death; we tell death to "Go to hell!" for that is where it must go.[76] Medical technology might numb the sting of death or postpone the inevitability of death. But Christ alone defeated death perfectly, fully, and eternally.[77]

Death is the curse of God spoken over creation after the fall. Like Adam hiding from God in the garden, man tries to hide from

70 John 2:19; Rom. 8:11; Gal. 1:1.
71 1 Cor. 15:12–28.
72 2 Cor. 4:14.
73 Col. 3:1–4.
74 2 Cor. 5:17.
75 Harari, *Homo Deus*, 23.
76 Rev. 20:14.
77 1 Cor. 15:55.

his inevitable end. But God has pronounced death for Adam's sin and for ours. How long will God let man hide? How long will he let mankind push back death before his attempts to brush against God's curse meet his final resistance? Will that be after we learn to live for 150 years, or two hundred years, or a birthday count closer to Methuselah? Maybe at some point anti-aging will push death back to the point of meeting the curse directly. After pushing aside "the inevitable" and extending mankind's life spans, perhaps we will reach an electrified fence, or maybe a flaming sword will finally block our path to immortality.[78] What if perfectly healthy bodies die for no other reason than running into God's divine limiter?

Universal death is the result of God's curse. But in Christ, physical death cannot win. Its power has been broken forever. When sinners like you and me look upon the Son, death's tyranny over our lives becomes a doorway into the resurrection and to an eternal existence that is tear-free, regret-free, and pain-free. This is the good news, the essence of Christianity, and the true gospel. Through Christ, our lives find joy and hope and balance. Through the gospel, we find clarity on all the gifts of this world—food, sex, money, and technology. Those things become true gifts, no longer gods.

Christ's victory over death in the first century is the key to our lives today. Because even if we eventually learn to suck the conscious part of us out of this decaying biohazard we call a body and upload it into a deathless digital cloud, what happens if we're stuck there with no eternal way out? What if a gnostic, body-less encapsulated existence is a sort of hell of its own? Or what if our technologies become so good that no one will die from "natural" causes, but only from violent accidents? This scenario would paralyze us all

78 Gen. 3:24.

into risk-less bubbles of inaction and make us "the most anxious people in history."[79]

Instead, as Lewis foretold, the overzealous impulse to save humanity is what destroys, because it comes from no higher ideal than trying to prolong life in this universe. His words apply to our new aspirations for Mars. "Nothing is more likely to destroy a species or a nation than a determination to survive at all costs," Lewis said. Instead, "those who care for something else more than civilization are the only people by whom civilization is at all likely to be preserved."[80] In Christ we overcome the fear of physical death. We, the grave conquers, are positioned to become the world's best humanitarians.

Lost in Space

The tech age will continue to ignore the resurrection and clamp human hope to rocket ships. Deep space is mesmerizing to the earthbound. The depth, width, and height of the universe is impossible to comprehend—at least 100 million galaxies stretching across at least ninety billion light-years. Neither number fits on my iPhone calculator. Neither amount makes any sense to my brain.

But for generations, people have looked up into the sky and claimed that as we reach out into these unfathomable expanses, we meet the absence of God. It's light-years of nothingness up there, they say, within this limitless cosmic expanse, said to exist for billions of years. So why does the Bible zoom in on one blue marble for just a few thousand years? That's the question many atheists ask—and answer by declaring that the earth-centric,

79 Harari, *Homo Deus*, 25.
80 C. S. Lewis, *Essay Collection and Other Short Pieces* (New York: HarperCollins, 2000), 365–66.

man-centric worldview of the Bible cannot hold up in the age of space exploration.

This is the conclusion of atheist philosopher Nicholas Everitt in his book *The Non-Existence of God*. There he claims, "Everything that modern science tells us about the size and scale and nature of the universe around us reveals it to be strikingly inapt as an expression of a set of divine intentions of the kind that theism postulates." Thus, "the findings of modern science *significantly reduce the probability* that theism is true, because the universe is turning out to be very unlike the sort of universe which we would have expected, had theism been true."[81] He claims that an earth-centric book like the Bible doesn't prepare us to discover 100 billion galaxies. Therefore, the Bible is made irrelevant in the age of space exploration.

Or as Carl Sagan said, "If you lived two or three millennia ago, there was no shame in holding that the Universe was made for us."[82] But you'd be stump-dumb to claim it now. Why? Because as our rockets push deeper into space and discover more of the expanse of our cosmos, our atheist astronomers postulate that, given the magnitude of the cosmos and the vast number of planets sure to exist among the far-flung galaxies, the earth is nothing special. More intelligent life surely exists somewhere else. We are not unique. And if we are not unique, any earth-focused deity must be shortsighted (at best), or the mythical product of some earthling's active imagination (at worst). These are strong speculations.

But after decades of ambitious photography and deep space probes, the same astronomer was led to make a more concrete con-

---

81 Nicholas Everitt, *The Non-Existence of God* (Oxfordshire, UK: Routledge, 2003), 218; emphasis original.

82 Carl Sagan, *Pale Blue Dot: A Vision of the Human Future in Space* (New York: Ballantine, 1997), 50.

clusion: "in terms of actual knowledge," there isn't another world "known to harbor a microbe, much less a technical civilization."[83] Study all the millions of deep-space photographs, and you'll find "no signs of a technical civilization reworking the surface of any of these worlds," Sagan concluded.[84] When the atheist scientist stops rhapsodizing his theories for a moment to check the documented facts (that is, the science), the evidence to date suggests that with every photograph of new, distant, vacant spheres in the vast beyond, earth is becoming more scientifically unique, not less. For now, there's no scientific reason to stop concluding that the entire cosmos was made as a theater for earthlings.

So scientist atheists tend to assume that it was Nicolaus Copernicus's orbital reorg that knocked humanity out of the center of the universe, followed by Sputnik, Voyager, and Apollo—which awakened to us the unfathomable scope of the cosmos and brought the ancient religions of man to the brink of existential crisis. Mankind is just a one-in-a-billion accident. But this line of thinking is too narrow, because if the God revealed in Scripture is true, we would expect to find his nature etched on the visible universe. And that's what we find. Space seems endlessly expansive to us, not because God is a myth, but because God has made himself unavoidably obvious to his creatures.[85] God's infinite vastness surrounds us in the metaphor of space.

Before low earth orbit satellites distracted our view of the greater cosmos, like gnats circling around your head, and before light pollution dimmed the glorious Milky Way, like a flashlight beam to your eyes, David stared into the radiance of deep space and

83 Sagan, *Pale Blue Dot*, 68.
84 Sagan, *Pale Blue Dot*, 120.
85 Rom. 1:18–32.

pondered: "Why would the God of all *that* pay attention to *me*?" It has always astonished mankind that the earth was chosen to be God's central stage in the drama of the cosmos. When I look up at night into the darkness and behold the glowing moon and distant planets and the five thousand flaming stars still visible to the naked eye; when I see all that you have made with your own fingers and set in orbit, I ask: "What is man that you are mindful of him, and the son of man that you care for him?" (see Ps. 8:3–4).

God created billions of galaxies, but the drama of history plays out on just one blue rock. So man pondered God's favor to us on this planet long before modern atheists used the expanse of the cosmos to suppress God in unbelief. Our universe has purpose beyond our perception. "The universe is not, certainly, exhausted by its service to humanity and must therefore have some goal other than utility to man."[86] Pragmatics cannot explain the scope of the known universe. It exists beyond what we can currently discover; it magnifies the power and worth of its Creator.

The far expanse of space is not a subjective accident, an unfinished afterthought, or a cosmic photobooth for satellites. Space is an objective, ordained, and undeniable revelation of God—"namely, his eternal power and divine nature" (Rom. 1:20). Theologically, space is never "to be considered as a 'fact of nature' whose meaning is self-evident without reference to the divine presence and claim upon creatures."[87]

So if we approach space exploration rightly, we don't meet the absence of God, but are welcomed into communion with the Creator. Many people, both Christians and non-Christians, struggle

86 Bavinck, *Reformed Dogmatics*, 2:433.
87 John Webster, *Confessing God: Essays in Christian Dogmatics II* (London: T&T Clark, 2005), 103–7.

to sense the presence of God in the massive expanses of space. But the Creator is infinite, and his infinity echoes in creation, including deep space. God made an open star cluster and named it "Pleiades." And he created a constellation of stars and named it "Orion."[88] He determined the exact number of all the combined stars in the universe and named each and every one of them.[89] Our latest guesses are that between 100 billion and two trillion galaxies exist beyond our present reach, and they all exist to teach us theology, to show God's covenant faithfulness to his people, and to declare the immensity and glory and love of the Creator.

Who Then Will We Trust?

Within God's universe, technology becomes meaningful for Christians who fear God and serve others. But we still live in a fallen world, with wars and rumors of wars and anger between people and nations.[90] No trip to the moon or mission to Mars can end human aggression. And the resulting war horses and war chariots are not entirely contrary to God's will, because God has sanctioned governments to lawfully wield swords and guns and tanks to punish and defend.[91]

We need technology, but we often idolize technology. The people of God were condemned in Isaiah because they turned to human discovery and innovation to find their security. When it comes to saving ourselves, we amass wealth and chariots, cling to idols, hide ourselves inside our innovations, climb up high towers, and crouch behind fortified walls.[92] When it comes to salvation, technology is

88 Amos 5:8.
89 Ps. 147:4.
90 Matt. 24:6; Mark 13:7.
91 Rom. 13:1–7.
92 Isa. 2:6–22.

an inevitable letdown. "The war horse is a false hope for salvation, and by its great might it cannot rescue" (Ps. 33:17). Man cannot save himself. "Stop regarding man in whose nostrils is breath, for of what account is he?" (Isa. 2:22). No innovator and no innovation can ultimately save you.

Science materializes our deepest heart longings into machines that we set our hopes on. "For humanity, technology is not simply a means to physical survival; it is also a means toward whatever he conceives of as spiritual fulfillment."[93] So we can agree with Paul Virilio, a critic of modern technology, when he says that we resist "not technology proper, but the propensity for us to confer upon technology a salvific function."[94] Indeed, "it would be unforgivable to allow ourselves to be deceived by the kind of utopia which insinuates that technology will ultimately bring about happiness and a greater sense of humanity."[95] It won't. Technology can do many things, but it will never satisfy our souls. Misused human innovation makes us spiritually numb. This is why the prophet Isaiah used brass and iron as metaphors for the spiritual dullness of sinful hearts.[96]

And yet what we make with brass and iron and silicon is amazing. The microprocessor is the most powerful thing on earth, and we own several of them without a license, permit, or government approval. The computers, smartphones, and gaming systems powered by these chips are incredible. The tiny flexible wires that could be inserted into the brain of a paralyzed man to grant him new

93 Victor C. Ferkiss, "Technology and the Future of Man," *Review and Expositor* 69, no. 1 (1972): 50.
94 Said of Paul Virilio in Gil Germain, *Spirits in the Material World* (Lanham, MO: Lexington, 2009), 102.
95 Paul Virilio, *Politics of the Very Worst* (New York: Semiotext(e), 1999), 79
96 Isa. 48:4.

telekinetic powers of motion are remarkable. The end of polio and the end of cancer and the end of dementia, should we see all these victories, would be astonishing feats of medical technology, and we would give God all the glory for them, because he creates minds to answer these problems and to push back against the fall.

But all these advances cannot fix what is truly broken.

The Soul

Computer scientist Alan Kay once joked that technology is anything invented after you were born.[97] But technology is far more expansive, and it includes the gifts of the nineteenth century that we now take for granted. Our innovation-saturated life today is a divine gift to us through the channel of many brilliant minds who were busy innovating decades before we were born.

The minds of the nineteenth century sped up tech advance. But as tech advances accelerated, contemporary preachers stepped into pulpits with concerns. With three-quarters of the century done, pastor J. C. Ryle cautioned that innovation blinds us to more important matters, saying: "We live in an age of progress—an age of steam-engines and machinery, of locomotion and invention. We live in an age when the multitude are increasingly absorbed in earthly things—in railways, and docks, and mines, and commerce, and trade, and banks, and shops, and cotton, and corn, and iron, and gold. We live in an age when there is a false glare on the things of time, and a great mist over the things of eternity."[98]

In this wave of modern marvels, Charles Spurgeon was wise enough to see human innovation as the harnessing of God's patterns in creation. But even as he celebrated new innovation, he offered

97 Kelly, *What Technology Wants*, 235.
98 J. C. Ryle, *Old Paths* (London: Charles J. Thynne, 1898), 41.

strong caution. "We have heard of engineers who could bridge the widest gulfs," he said from the pulpit. "We have seen men who could force the lightning's flash to carry a message for them; we know that men can control the sunbeams for their photography, and electricity for their telegraphy; but where dwells the man, where even is the angel, who can convert an immortal soul?"[99] Technology is powerless to turn sinners away from the false promises of this world in order to find their satisfaction in Christ. Regeneration is the truly remarkable power in any age.

God deposited a wealth of innovations and technological potential into creation, not to lead us into temptation, but to reveal what we most love and where we place our greatest trust. That may be in God. Or that may be in chariots, missiles, cyberwar, vaccines, dataism, AI, or anti-aging discoveries. "Some trust in chariots and some in horses, but we trust in the name of the LORD our God" (Ps. 20:7). This is the human dilemma. Human innovation is a wonderful gift but a disappointing god. We cannot save ourselves. In the end, our innovations leave hearts unsatisfied, souls lost, and bodies cold in the grave.

99 C. H. Spurgeon, *C. H. Spurgeon's Forgotten Early Sermons: A Companion to the New Park Street Pulpit: Twenty-Eight Sermons Compiled from the Sword and the Trowel* (Leominster, UK: Day One, 2010), 258–59.

5

When Do Our Technologies End?

WHENEVER I SPEAK OF the Gospel of Technology as a single, coalescing movement (as I implied in the previous chapter), I expect disagreement. For tech optimists like me, this seems too pessimistic. It is reasonable to address scientific problems in a given industry, like genetics or AI or autonomous bots. But collecting them all into one Gospel of Technology goes too far. It fails to weed the good from the bad. People will always abuse technology, yes. But it is not a unified conspiracy. So is it overboard to suggest that the tech industry is coalescing religiously and colluding into one massive anti-gospel?

In reality, man has always loved to innovate new techno-religions in the spirit of Babel. A fair number of historians suggest that ancient religions emerged as ways of appeasing the gods in order to control nature. But now that we can control our environments with technology, we don't need the ancient religions. Technological control displaces our gods. Or, maybe more accurately, technology becomes the culture's new religion. Maybe, in the words of computer philosopher Jaron Lanier, "what we are seeing is a new religion, expressed through an engineering culture."[1] Thus, the engineered religions of the future will not come from the Holy Land, writes Yuval Noah Harari; "they will emerge from research laboratories. Just as socialism took over the world by promising salvation through steam and electricity, so in the coming decades new techno-religions may conquer the world by promising salvation through algorithms and genes." In other words, the Holy City of the future is not Jerusalem, but a technopolis like Silicon Valley. "That's where hi-tech gurus are brewing for us brave new religions that have little to do with God, and everything to do with technology. They promise all the old prizes—happiness, peace, prosperity, and even eternal life—but here on earth with the help of technology, rather than after death with the help of celestial beings."[2] Tech engineers are emerging as today's priestly class, not unlike the ancient blacksmith in previous generations.

So we must turn our focus to the future, not because our tech gurus point there but because Scripture does. The Bible reveals to us a future moment when God's relationship to technology takes a decisive turn. Over time, Scripture foretells that our technology will take on a redemptive role in a reunified civilization that seeks happiness, peace, prosperity, and eternal life by its own inventions.

If this sounds familiar, it is. This spirit of idolatry emerged in Babel, a primitive city-tower, an altar built to a new religion engineered by man to celebrate man. Then God stepped into the story— stepped *down* into the story—to judge man, not with extermination, but with confusion. God thwarted human "progress" by multiplying the languages of earth and scattering hundreds of unique cultures all across the globe. This act was not the end of the city but the genesis

1   Jaron Lanier, *Who Owns the Future?* (New York: Simon & Schuster, 2014), 193.
2   Yuval Noah Harari, *Homo Deus: A Brief History of Tomorrow* (New York: HarperCollins, 2017), 356.

of a thousand cities. It was a temporary mercy. God was fully aware that this dispersion of cultures would eventually undo itself and Babel would recollect and return in the form of man's greatest city, Babylon—or as she preferred to be called, "Babylon the Great."

Sex and the City

Babylon is super-sized Babel, a godless city noted for its desire to exalt herself above God. Babylon is the culmination of all cities, a picture at the end of our Bibles of all man's cities fully realized. Tokyo, Delhi, Shanghai, Cairo, Beijing, New York, Istanbul, Moscow—Babylon is the ultimate city, a composite of every city, the highest expression of man's urban dreams and aspirations.

In Babylon are "all cities condemned."[3] Those are the blunt words of Jacques Ellul in his book *The Meaning of the City*, a survey of God's tumultuous relationship with the cities of man. The story is often ugly, as we saw back in Babel. But Ellul says that the origin of the city begins with goodwill. In the city, "everyone works to enable man to live better." The city offers access to better homes, food, services, culture, and entertainment to occupy us as we resist the dreariness of life. The city connects us with others to escape loneliness. It offers us networks and shared skills and better jobs. It buffers us from the seasons, shields us from foreign invasion, and protects us with close access to the latest emergency medicine. Cities give us "more comfort and what are called the joys of life, with all the guarantees of science, medicine, and pharmacology at his doorstep. To change the powerlessness of him who must watch those he loves die, unable to do a thing."[4] The medical tech of the city pushes back death. It is where many

3 Jacques Ellul, *The Meaning of the City* (Eugene, OR: Wipf & Stock, 2011), 49.
4 Ellul, *Meaning of the City*, 60.

people apply their skills in service of others. Ideally, the city is a place of love.

But when love for God fails to take hold at the center of a society, the resulting humanitarianism falls under the umbrella of what Reno calls "the deadly tyranny of philanthropy." The highest aspiration of the city eventually ends in "rallying all the forces at our disposal to serve whatever god of worldly flourishing we have made for ourselves."[5] Concerned with helping mankind flourish apart from God, cities impose apparent "goods" on others. Every well-meaning city, focused on the needs of man but ignoring the glory of God, will eventually become a dehumanizing place, the kind of place where the false promises of the Gospel of Technology proliferate in the name of humanitarianism.

This is the story of "Babylon the Great" and her overthrow. Revelation 18 tells the story, as the apostle John watches the final events of history unfold in events leading up to the return of Christ to earth. We begin with verses 1–3.

[1] After this I saw another angel coming down from heaven, having great authority, and the earth was made bright with his glory. [2] And he called out with a mighty voice,

> "Fallen, fallen is Babylon the great!
>   She has become a dwelling place for demons,
> a haunt for every unclean spirit,
>   a haunt for every unclean bird,
>   a haunt for every unclean and detestable beast.
> [3] For all nations have drunk

---

5   R. R. Reno, *Genesis* (Grand Rapids, MI: Brazos Press, 2010), 133.

the wine of the passion of her sexual immorality,
and the kings of the earth have committed immorality
with her,
and the merchants of the earth have grown rich from the
power of her luxurious living."

Babylon is a global epicenter of wealth, opulence, comfort, consumption, war machines, and technology. Cities compress human activity. Cities are technologies of their own, functioning like microprocessors, "a wonderful technological invention that concentrates the flow of energy and minds into computer chip–like density. In a relatively small footprint, a city not only provides living quarters and occupations in a minimum of space, but also generates a maximum of ideas and inventions."[6] Cities concentrate both human innovation and human arrogance, obvious in Babel and now in Babylon. Cities are capitals of self-sufficiency, of "the confidence of human beings that they could find security through their own technological expertise."[7]

Inside cities, technological advances conspire to evict God. God has become irrelevant. The Babylonians have grown so confident in their skill to meet every problem that they view the Creator as irrelevant to the drama of human enterprise.[8] God is unnecessary. Babylon is the global capital of self-sufficiency, filled with drunk idolaters and opulent adulterers, a city that has "deserted God and replaced him with other lovers."[9] Babylon is her own god. She is the summit that every other man-made megacity aspires to reach. As a

6 Kevin Kelly, *What Technology Wants* (New York: Penguin, 2011), 84.
7 J. A. Motyer, *The Prophecy of Isaiah: An Introduction and Commentary* (Downers Grove, IL: InterVarsity Press, 1996), 176.
8 T. Desmond Alexander, *The City of God and the Goal of Creation* (Wheaton, IL: Crossway, 2018), 28–29.
9 Alexander, *City of God and the Goal of Creation*, 145.

result, urban centers ignore God's beauty for a vast global network of misdirected affections. Self-worshiping sinners waste their lives on a vain quest to find satisfaction in wealth, sex, and power.

Reminiscent of Babel, God turns his attention to Babylon to thwart its self-sufficient spirit. But first he calls his faithful to exit the city.

⁴ Then I heard another voice from heaven saying,

> "Come out of her, my people,
>> lest you take part in her sins,
> lest you share in her plagues;
> ⁵ for her sins are heaped high as heaven,
>> and God has remembered her iniquities.
> ⁶ Pay her back as she herself has paid back others,
>> and repay her double for her deeds;
>> mix a double portion for her in the cup she mixed.
> ⁷ As she glorified herself and lived in luxury,
>> so give her a like measure of torment and mourning,
> since in her heart she says,
>> 'I sit as a queen,
> I am no widow,
>> and mourning I shall never see.'"

The eyes of the world will turn toward the city, but the eyes of God's people will turn away. God's people will depart from the city, summoned by divine exodus. More on this in a moment.

The spirit of Babylon is the spirit of transhumanism. Made arrogant through opulence and comfort, Babylon now claims power over the grave. "She vaunts her power, even against death. She needs

WHEN DO OUR TECHNOLOGIES END?

no one. Her base of power is no mere human being. She is her own reason for existing, in herself a sufficient power, a sufficient law. She excludes God because she is for herself her own sufficient spirituality."[10] She thinks that she overcame death and sorrow; she thinks that she usurped God. Babylon seduces the world with promises of joy and security. Babylon's promise of immortality is the Gospel of Technology's promise today.

And yet in this transhuman lust for victory over death and for self-discovered immortality, the rebel city only stokes the fire of her coming judgment, which will consume her arrogant God-rejection with literal and eternal death.

> [8] "For this reason her plagues will come in a single day,
>   death and mourning and famine,
> and she will be burned up with fire;
>   for mighty is the Lord God who has judged her."

> [9] And the kings of the earth, who committed sexual immorality and lived in luxury with her, will weep and wail over her when they see the smoke of her burning. [10] They will stand far off, in fear of her torment, and say,

> "Alas! Alas! You great city,
>   you mighty city, Babylon!
> For in a single hour your judgment has come."

Babylon will burn to the ground. In a flash, in a single-day fire, the mighty city will burn as a global spectacle of smoke for the eyes of

10 Ellul, *Meaning of the City*, 52.

all the kings of earth who fed the opulence and sinful indulgence of the city. Babylon will boast in no final solution from these fires and famines. It will reap the judgment of her idolatries, as the world's merchants watch their profits burn.

> [11]And the merchants of the earth weep and mourn for her, since no one buys their cargo anymore, [12]cargo of gold, silver, jewels, pearls, fine linen, purple cloth, silk, scarlet cloth, all kinds of scented wood, all kinds of articles of ivory, all kinds of articles of costly wood, bronze, iron and marble, [13]cinnamon, spice, incense, myrrh, frankincense, wine, oil, fine flour, wheat, cattle and sheep, horses and chariots, and slaves, that is, human souls.

Babylon's ports welcomed daily necessities like oil, flour, and wheat. They also welcomed an abundance of animals for breeding and eating. But the city was especially drawn to imported luxury goods from across the world: expensive jewelry and clothing, exotic spices and fragrances, and the best home furnishings the world offered. So also flowed a stream of iron for metal tools and weapons, marble for awe-inspiring building projects, and horses and chariots to feed the city's lust for entertainment.[11]

People end the list on purpose, as expendable merchandise. Cities consume high amounts of energy, and before electricity, cities ran on imported humans. "Slavery was to the ancient world, more or less, what steam, oil, gas, electricity and nuclear power are to the modern world. Slavery was how things got done."[12] The wealth and opulence of Babylon was built on the backs of

---

11   Ian Boxall, *The Revelation of Saint John*, Black's New Testament Commentary (London: Continuum, 2006), 260–61.

12   N. T. Wright, *Revelation for Everyone* (Louisville, KY: Westminster John Knox, 2011), 164.

bodies who are eternal souls, worshiping beings, used simply for their raw power. Babylon dehumanized God's image bearers into animals, functional commodities bought and sold like cattle, sheep, and horses.

On their ships, the merchants witness the destruction and are brought to tears. Babylon is the heart of a global economic empire, and its destruction means the end of an idolatrous, global market.

> [14] "The fruit for which your soul longed
>   has gone from you,
> and all your delicacies and your splendors
>   are lost to you,
>   never to be found again!"

[15] The merchants of these wares, who gained wealth from her, will stand far off, in fear of her torment, weeping and mourning aloud,

> [16] "Alas, alas, for the great city
>   that was clothed in fine linen,
>   in purple and scarlet,
> adorned with gold,
>   with jewels, and with pearls!
> [17] For in a single hour all this wealth has been laid waste."

And all shipmasters and seafaring men, sailors and all whose trade is on the sea, stood far off [18] and cried out as they saw the smoke of her burning,

"What city was like the great city?"

[19] And they threw dust on their heads as they wept and mourned, crying out,

> "Alas, alas, for the great city
>   where all who had ships at sea
>   grew rich by her wealth!
> For in a single hour she has been laid waste.
> [20] Rejoice over her, O heaven,
>   and you saints and apostles and prophets,
> for God has given judgment for you against her!"

Then comes a moment of theatric subterfuge. Babylon, with all her opulence and power and innovation and oppression, shall be like a millstone thrown into the ocean.

> [21] Then a mighty angel took up a stone like a great millstone and threw it into the sea, saying,

> "So will Babylon the great city be thrown down with
>   violence,
> and will be found no more."

Paradoxically, Babylon's mills, the icon of her ancient industrial power, and a technique taught to man by God for mass-producing flour, will stop grinding. The idolatry is stopped by a rock the size of a two-ton millstone. The doom of the great city is matched to her industrial strength. Her judgment emerges from creation to stop the tools that she made from creation. The loud sound of this industrial and technological system, built out of defiance of God, will be silenced and drowned. Babylon is cast to the bottom of the ocean. Her economy drowns, her industry halts, her music stops, and her innovators build no more.

> [22] And the sound of harpists and musicians, of flute players
>   and trumpeters,
>   will be heard in you no more,
> and a craftsman of any craft
>   will be found in you no more,
> and the sound of the mill
>   will be heard in you no more.

Babylon is where Jabal and Jubal and Tubal-cain's innovations found their highest expression, but those innovations are now ended. The blacksmith's anvil rusts. Global shipping, commerce, imports and exports cease. Music stops, and "without music, civic life grinds to a halt."[13] As the social soundtrack is quieted, so too are the reverberations of manufacturing. The heavy millstones, once the loud grinding sound of the economic machine of commerce, are silenced. There is no music. No industry. No imports. No tool making. No more grain to feed man or livestock. The final end of Babylon in Revelation 18:22 is the bookend to Genesis 4:19–22. Cain's lineage is finally undone. All the godless wielders of technology, who had grown wealthy in their industries, are ended.

God gave music making, tool making, and cattle breeding to man, but not for man to use for whatever selfish purposes he has for wealth and power and opulence. So before the eyes of the world, the global industrial-economic system of human wealth burns to ash—a fiery end to the legacies of Cain's children and to Cain's

13 Peter J. Leithart, *Revelation*, International Theological Commentary on the Holy Scripture of the Old and New Testaments (London: T&T Clark, 2018), 2:245.

heritage as a city builder. The brightest human city, the greatest metropolis in history, goes dark. Its culture making ends.

> [23] And the light of a lamp
>   will shine in you no more,
> and the voice of bridegroom and bride
>   will be heard in you no more,
> for your merchants were the great ones of the earth,
>   and all nations were deceived by your sorcery.
> [24] And in her was found the blood of prophets and of saints,
>   and of all who have been slain on earth.

Babylon actively persecuted Christians to the point of bloodshed. Because Christians rejected the idols of the city, they were unwelcome in the marketplace. "Babylon's economic system persecuted Christian communities by ostracizing from the various trade guilds those who did not conform to worship of the guilds' patron deities. This usually resulted in loss of economic standing and poverty for those ostracized. In a real sense, this meant the removal of Christian artisans from the marketplace and a removal of the common pleasures of life enjoyed in normal economic times."[14] By refusing to participate in the sexual idolatry of the city or to worship its opulence, Christians were shunned by the market.

Any society that worships the Gospel of Technology will discriminate against the gospel of Jesus Christ. Our ancient priorities look foolish to the tech culture around us. Even today, corporate structures can make Christians feel that they are out of place, that their faith is unwelcome, and that they are safer if they keep their

---

14 G. K. Beale, *The Book of Revelation: A Commentary on the Greek Text,* New International Greek Testament Commentary (Grand Rapids, MI: Eerdmans, 1999), 919.

convictions to themselves. Christians may be silenced and eventually shut out of the culture's major publishing and social-media platforms for speaking the truth. This is not postmodern. This is Babylonian. And as in Babylon, in extreme cases, martyrs die. But such an urban tragedy is a form of urban resistance too. For Christians inside innovation-driven cities like Babylon, martyrdom showcases "the power-in-weakness of the kingdom of God rather than the power-knowledge of earthly kingdoms."[15]

In the end, Babylon says that the purpose of humanity is to glorify oneself and enjoy self-sufficient opulence and idolatry forever. But the true purpose of humanity is to glorify God by enjoying him forever. "To focus on humanity as the center of everything and to forget God is the greatest sin."[16] Self-glorification calls down God's justice. This was the sin of Babel and the sin of Babylon. This is the sin of the Gospel of Technology.

Takeaways

Whether brought to the point of bloodshed or not, technically advanced urban centers pressure Christians to distance themselves from hope in Christ. Cities may try to silence Christianity, but God has the last word. The smoldering end of man's largest city leaves us with a few takeaways that city dwellers need today.

1. The city is the engine of innovation.

God created three heroic innovators to preserve Cain's lineage: Jabal, Jubal, and Tubal-cain. Ever since, human innovation has been closely tied to the city. And while the urban planning of man

---

15 Kevin J. Vanhoozer, *Pictures at a Theological Exhibition: Scenes of the Church's Worship, Witness and Wisdom* (Downers Grove, IL: IVP Academic, 2016), 280.
16 Beale, *Book of Revelation*, 921–22.

could only operate within motives of idolatry and pride and greed, the city remained a merciful gift. We build a diversity of cities as we are led by our cultural impulses and as we are taught by the Creator to follow his creational patterns. It is a great mercy from God to find your way into the safety of a city.[17]

Jacques Ellul, on the other hand, roots the genesis of all city building in man's fundamentally rebellious attempt to concoct a false Eden. Man raises up a fortress of safety behind steel and concrete, a work of countercreation, a diabolical counterfeit, a habitat of artifice, a fraudulent rejection of God's creation in order to replace it with man's creation. God does not stop this rebellion but adopts it into his own plan for Old Jerusalem and New Jerusalem.[18] This is too far. Ellul fails to appreciate God's plan for preserving Cain and his lineage, the forefathers of the city.

On the contrary, we can appreciate cities for what they are—temporary gifts from God for good purposes, particularly to serve as epicenters of innovation. God alone gifts a city with wealth and well-being, and from these gifts come new possibilities of science and technology.[19] Without wealth and health, cities cannot innovate. But with these blessings in place (and they are only put in place by God's hand), the gears of technological development turn and networks form. "The city features street life and marketplaces, bringing about more person-to-person interactions and exchanges in a day than are possible anywhere else," writes Tim Keller, suggesting again that the city is something like a microprocessor. "The more often people of the same profession come together, the more

---

17 Ps. 107:1–9.

18 Ellul, *Meaning of the City*, 77, 102–3.

19 Herman Bavinck, "Common Grace," trans. Raymond C. Van Leeuwen, *Calvin Theological Journal* 24 (1988): 60.

they stimulate new ideas and the faster these new ideas spread. The greater the supply of talent, the greater the productivity of that talent, and the demand for it follows."[20] The tech age flourishes because of cities. This is good and bad. It means that the Gospel of Technology will be conceived and nurtured inside the womb of the city. But it also means that from within the confines of Cain's city, new beneficial innovations will emerge and serve each of our lives with health, tools, and culture.

2. Christians integrate into cities.

So if the technium will eventually turn against believers, should Christians exclude themselves from the technium of Babylon? This is one of the deep challenges of life in the technological age. It's the conundrum of the city, because Babylon is not just a city—she's every city.

But she's also an ancient city. Future Babylon should remind us of original Babylon, an ancient city led by its infamous ruler Nebuchadnezzar. Ancient Babylon was so culturally advanced that it now provides "the basis upon which our culture is built."[21] This same ancient city was a dire threat to God's people in Jerusalem. And when Nebuchadnezzar de-tech-ed the holy city by exiling its war technicians and metalworkers, Jerusalem was rendered defenseless to Babylon's takeover.[22]

In exile under Nebuchadnezzar, what were God's people called to do? Were they called to revolt against the wickedness of this pagan city, and escape like the exodus from Egypt? No. God had

20 Timothy Keller, *Center Church: Doing Balanced, Gospel-Centered Ministry in Your City* (Grand Rapids, MI: Zondervan, 2012), 137–38.
21 Herman Bavinck, *The Wonderful Works of God* (Glenside, PA: Westminster Seminary Press, 2019), 39.
22 Jer. 29:2.

a different plan this time. He did not call his people to run and escape. He did not enact a second exodus for them, at least not immediately. And he didn't tell them to become a cancer to destroy Babylon from the inside. He called them to the exact opposite response. The prophet Jeremiah communicated God's astonishing commands:

> 4 "Thus says the LORD of hosts, the God of Israel, to all the exiles whom I have sent into exile from Jerusalem to Babylon: 5 Build houses and live in them; plant gardens and eat their produce. 6 Take wives and have sons and daughters; take wives for your sons, and give your daughters in marriage, that they may bear sons and daughters; multiply there, and do not decrease. 7 But seek the welfare of the city where I have sent you into exile, and pray to the LORD on its behalf, for in its welfare you will find your welfare." . . . 10 "For thus says the LORD: When seventy years are completed for Babylon, I will visit you, and I will fulfill to you my promise and bring you back to this place. 11 For I know the plans I have for you, declares the LORD, plans for welfare and not for evil, to give you a future and a hope." (Jer. 29:4–7, 10–11)

God's people were called to seek the welfare of the advanced pagan city. From the inside. This principle is immediately relevant to all city-dwelling Christians today. Incredibly, "our job is to lead the life of the other inhabitants of the city," writes Ellul. "We are to build houses, marry, have children. What a happy ground for conciliation, for that is exactly what the city is asking of us!" And so we are called "to continue from one generation to the next" by contributing to the "very stability and depth which men were seek-

WHEN DO OUR TECHNOLOGIES END?

ing when they built the cities." God's people are not city makers; we are city dwellers. We are not called to make cities, but to make the cities we live in better. We avoid the false spiritual promises that animate our cities, but at the same time we are "clearly told to participate materially in the life of the city and to foster its welfare." We are called "to share in the prosperity of the city, do business in it, and increase its population," even to "defend it because our solidarity is there."[23]

God's people in ancient Babylon "are called to be the *very best* residents of this particular city of man. God commands the Jewish exiles not to attack, despise, or flee the city—but to seek its peace, to love the city as they grow in numbers."[24] Even within a wicked city, the people of God seek to flourish in their callings.

3. *Christians leaven the city before fleeing the city.*

Since Sodom, cities have been synonymous with sexual sin. And in the case of Sodom, God decided to end it with fire. But before he did, he showed us a lesson. Abraham, the one who evaded cities, pleaded on behalf of God to spare the people of the city. Should fifty righteous believers live in the city, would God preserve it? Yes, God said. For fifty righteous believers the city would not be destroyed. Abraham persisted. What about forty-five, or thirty, or twenty, or ten? Same answer. God would not destroy a wicked city for the sake of ten believers within it.[25]

What can we learn here? At the end of time, God will separate the wicked and the righteous, but he does not separate them in the city. Until then, Christians leaven their cities. Ellul writes, "The

23 Ellul, *Meaning of the City*, 74.
24 Keller, *Center Church*, 143.
25 Gen. 18:22–33.

entire city is spared when there is one pocket of righteousness, no matter how feeble, hidden in her midst. And this opens up a possibility for the inhabitants to save their city. Not to save it from the last judgment, not from the univocal condemnation pronounced against the city, but from its execution here and now, on their particular city, on them its inhabitants, from that execution serving as notification of the final judgment."[26]

As we await the final reckoning, each metropolitan church will require a uniquely balanced diet of stern spiritual warnings and radiant eternal promises.[27] But that's the point. Despite the dangers, Christians will stay in cities to raise families and fulfill callings and enjoy the benefits of the city while they also learn to resist the God-less attitudes of the city. Just as every technology has its own inherent biases—Twitter celebrates sarcasm, Facebook celebrates fringe thinking, and Instagram celebrates bodies—every city has its own biases too. Each of the seven churches in Revelation battled their own spiritual idols. Each city today has its own idolatrous tendencies. Those biases do not drive Christians out of cities; it drives Christians to discernment within them.

In this way, a biblical theology of technology is simply a biblical theology of the city. The Bible's storyline of the city moves from Cain's first urban project, then to Babel's tower, then to Babylon's opulence, and ends with God's descending city. From the early chapters of Genesis to the later chapters of Revelation, the story of human innovation is braided into the story of city building. The challenges of embracing city life are the challenges of embracing tech culture. So if your conscience approves of living inside a city—among all of its cultural pressures and idolatrous biases—you

26 Ellul, *Meaning of the City*, 64.
27 Rev. 2:1–3:22.

are simultaneously preapproved to adopt new tech, even to work inside the tech industry.

But there will come a time when Christians are called out of Babylon. Ellul points to the beginning of our text in Revelation 18 and shows how one angel declares Babylon to be fallen before another angel arrives to call for God's children to remove themselves from the city.[28] "Thus the order to leave the city, to separate from her, is given when the city is already fallen, destroyed, when there is nothing else to be done to preserve and save her," Ellul writes. "When her judgment has been executed, and when, therefore, the Christian's role in her midst has no more meaning. It is this command from God which we must await." We await the call to leave the city as we go on living fruitfully inside the city. "How much easier it would be to reject the city now, to refuse her our presence now. But that cannot take place before God's final decision. And so we are involved in her life to the very last minute, and it is not in our power to disengage ourselves."[29]

Imagine a whole generation of Christians in exodus, called to put down their phones, close their computers, ignore the SUV in the garage, and walk out the front door of a home filled with comforts and tools, and evacuate their city on foot without looking back. Maybe this will be our generation, or the next generation, or the generation after that one. At some point Christians will heed the angelic decree to leave the city of man with all its wealth and power and gifts and blessings. The Giver will call us away from the wonderful, temporary gifts we enjoyed every day inside the city. Whether or not that is our lot, Babylon's future is meant to change us now. We live inside our cities and enjoy our gifts of tech, yet we

28 Rev. 18:2–4.
29 Ellul, *Meaning of the City*, 78–79.

live as men and women awaiting our exodus to a greater city. We hold our cities—like our gadgets—loosely.

A total tech disengagement is coming, but not yet.

God destroys cities (like Sodom). And he revives cities (like Nineveh). Whatever his plans for any city, we live inside them as leaven. We will be called from the city one day. But until then we thrive inside the city, even as we resist the spirit of the city in our hope and by our worship.

4. Christians resist the spirit of the city in their hope.

The Christian is never at ease, never really at home, inside the cities of man. The city's tech will never become a utopia. We live in the city today, but we hope for the city to come. That hope leaves us unsettled in man-made cities—just as our forefathers were unsettled and repelled from cities. By faith Abraham, Isaac, and Jacob lived in tents because they could see the false promises of the cities of man and were "looking forward to the city that has foundations, whose designer and builder is God" (Heb. 11:9–10). Even if your home is now rural, your hope is the same, the promise of gathering in God's holy city.[30]

Our faithful hope contrasts the hopeless city. The man-made city captures hearts and minds and imagines a man-made life severed from God. The Gospel of Technology flourishes in cities. Cities become a God-less utopian substitute for heaven, something like Babel. Yet we pray for God to spare our cities a little while longer. We pray for the Spirit to work in our cities. We pray for revival. But even as we love our cities, we "will be looked upon as adversaries of public welfare or as enemies of the human race and our efforts for the good of the

30 Jer. 29:14.

city will be interpreted as a will to destroy it."[31] That is not our aim, but it may become our accusation. Our waiting is subversive, but not against the city—we want other sinners to await the greater city too. We cling to the hope of a long-promised city, a new city made by God himself, and our hope is a direct affront to the transhumanists and the posthumanists and to all seekers of a man-made utopia. "Our waiting attitude, if it is constant and true, if it reaches our very hearts, is the very ruin of the spiritual power of the city."[32]

5. Christians resist the spirit of the city by their worship.

The exact contours of the future Babylon are malleable. But that future city will always coincide with the ancient city of Babylon. Ancient Babylon's hanging gardens were one of the wonders of innovation in the ancient world, "and represented human ingenuity at its best. Today Babylon is anti-God Western society with its glamour, glitz, technology, and entertainment that promises so much pleasure."[33] "Babylon is allegorical of the idolatry that any nation commits when it elevates material abundance, military prowess, technological sophistication, imperial grandeur, racial pride, and any other glorification of the creature over the Creator."[34] As we make sense of an ancient prophecy, written prior to electricity, it is no stretch to overlay technological advances into this affluent and arrogant city to come. The spirit of apocalyptic Babylon is the climate where the Gospel of Technology will incubate, grow, and rule over minds and hearts.

---

31 Ellul, *Meaning of the City*, 76.
32 Ellul, *Meaning of the City*, 78.
33 Anthony R. Petterson, *Haggai, Zechariah, and Malachi* (Downers Grove, IL: IVP Academic, 2015), 172.
34 Bruce Manning Metzger, *Breaking the Code: Understanding the Book of Revelation* (Nashville, TN: Abingdon, 1999), 88.

Fallen man's first reach for autonomy from God in the engineering of Babel will recollect into the greater technologies and innovation in Babylon. They're coming. But until we await God's destruction of Babylon, and as we await our great disengagement, how do we now resist the entire industrial-technological economic power complex, shown in Revelation 18? We follow the example in Revelation 19, as we worship the one true God. In the face of whatever superpower is to come, or whatever new discoveries are made in a lab, our calling never moves from worship. "The God of the Bible is also the God of the genome. He can be worshiped in the cathedral or in the laboratory."[35] Our worship "has nothing to do with pietistic retreat from the public world. It is the source of resistance to the idolatries of the public world."[36] Christian worship is public resistance. In the age of technologically derived idols, we are not called to preempt every evil use of technology.

We are not called to understand all technologies and their uses. Nor must we retreat from technological culture. We live within it. And we are called to live in such a way that we remind the world of what they never want to hear. We are called to point other sinners to the sole cause of all human possibility, the mind that patterned all of creation, the very maker of every innovator who seeks to satisfy our hearts with himself. More than our direct critique, our worship is a siren to the world to turn away from the worship of innovation and its idolatrous autonomy from the Creator. By our worship we deflate the arrogance of the technologist and show him that all

35 Francis S. Collins, *The Language of God: A Scientist Presents Evidence for Belief* (New York: Free Press, 2007), 211.

36 Richard Bauckham, *The Theology of the Book of Revelation* (New York: Cambridge University Press, 1993), 161.

innovation is made possible by a gracious and infinite Creator. All our making is derivative. Only God could make possible the city, with all her gold and wealth and innovation. God makes the city makers. No innovator can escape the domain of their Maker, since he created them (*bara*) to create, making people for both honorable uses and for dishonorable uses.[37]

Our worship cannot stop Babylon, but it will threaten her to the point of bloodshed. Babylon will kill our prophets and slaughter our saints, but Babylon will never stop our worship.[38] Our worship will perpetually remind Babylon's technologists that the living God of the universe is irrefutably sovereign and graciously merciful. As we near the end of this chapter and turn toward the ethics of life in the technological age, this point is crucial. In the age of innovation, the church stands firm, worshiping the one true God, exposing the false Gospel of Technology, and exulting in the joyful hope of eternity.

The Conclusion (Or New Beginning)

Human innovation and industry and medicine and music and genetics and commerce and astronautics all meet their end in Babylon. None of these fields of innovation live up to their full potential—none of them become a means to worship the Creator of all possibilities. They ultimately do the opposite. Science is infatuated by power, commerce is gorged by greed, and music is obsessed by worldliness. They are all diseased by a Babel-like self-confidence and a Babylon-like fixation on power and opulence. In this fallen world, scientists and technologists accomplish incredible feats made possible in creation but refuse to worship the Creator.

37 Rom. 9:21.
38 Rev. 18:24.

In the end, the whole city complex must be drowned and buried to make room for something better.

Babel must be razed to be replaced by a rival city. Indeed, the end of Cain and the desolation of Babylon were necessary to open the way for God to enter and to dwell in the new creation. The worship of God, done by a handful of souls inside a technologically autonomous metropolis, was a temporary concession. The worship of God was always intended to radiate out from the city center. And one day it will. Babylon will be thrown into the sea to make way for the New Jerusalem. Entering into the story is a new city not built by man but built by God. This city radiates with life and holiness and the presence of Christ, a temple-less city, "for its temple is the Lord God the Almighty and the Lamb." The new creation will need no solar panels, no LED light bulbs, no smartphone flashlights. It will not even need the sun or the moon in the sky, for the radiance of God's inexpressible light will provide all of its illumination.[39]

Like Babylon, God's eternal city will be the heart of a global network. All the nations will walk by the light of Christ, and by his light all the earth's kings will bring "the glory and the honor of the nations" into the city. Nothing unclean or selfish or self-glorifying will enter God's city.[40] When Babylon is destroyed, so too will be dashed all urban attempts to thwart God. There will be no basis to distrust cities, like in the pessimism of our forefathers Abraham, Isaac, and Jacob. This city will not be built by the design and ingenuity of man, like Jabal, Jubal, and Tubal-cain. Exceeding every known city in history, the New Jerusalem will arrive as the first city ever designed, engineered, and built by God himself.

39 Rev. 21:22–23.
40 Rev. 21:9–27.

Excursus: Old Tech in the New Creation?

At the end of the Bible's storyline we meet the destruction of man's great city and the arrival of God's city. And if these future realities are true, we meet another tech question: Will our familiar innovations in this life carry over into the new creation? How we answer that question, and how we understand technology in the New Jerusalem, depends on how we interpret two Bible texts.

The first passage regards the glory of the New Jerusalem. We are told that into God's city will flow, without end, "the glory and the honor of the nations."[a] Zion will be the global epicenter of worship. But will this glory include the wealth and discoveries and innovations of the nations? Or is it limited to human worship and praise?

The second passage prompts us to choose a literal or metaphorical interpretation of Jesus's declaration to his disciples. He told them: "If you have faith like a grain of mustard seed, you will say to this mountain, 'Move from here to there,' and it will move, and nothing will be impossible for you.'"[b] Should we take a literal or nonliteral interpretation here?

Here's how the options shake out, simplified down into two answers.

Position 1 (Spiritual/Literal):
Technology Ends in the New Jerusalem

In this view, "the glory and the honor of the nations" is limited to spiritual worship and praise. It has little to do with financial wealth or innovation. Eternity is tech-less because it is wealth-less. There's no buying and selling in eternity, no market economy as we know it today, and no need for technological development. Instead, everything returns to garden-like-ness. Wild food will grow abundantly, and if we grow crops, it will be with ease.

Farming and all our work will become labor-less. If we take a literal reading of Jesus's words to his disciples, God's children will spread across the globe to govern the new creation by willpower, without the mediating tools that we need under the curse today. Instead, in the resurrection, we will regain a Jedi-like power to cause creation to bend and move and respond according to our volition alone. In eternity, we will speak and the new creation will heed us, as it does for God. God's working and resting are essentially the same experience because nothing can resist his sovereign will. But for us in this life, especially after the fall, many forces conspire to resist our work. For us, *work* and *rest* are opposites—one pushing against resistance and the other succumbing to it.

Perhaps in an uncursed creation, freed from the defiance of thorns and thistles and dust and pain and sweat, we too will work in a resistance-less state closer to what we now call rest. Technology will give way to volitional willpower. With this literal interpretation, many Christians conclude that our technologies serve a temporary purpose in this age, to help us arm-wrestle the resistance of the curse.[c]

Position 2 (Material/Metaphorical):
Technology Continues into the New Jerusalem

In this second position, "the glory and the honor of the nations" includes global worship but is not limited to it. Interpreted through the prophecy of Isaiah 60, the glory must include a more metalic fruit in "the *wealth* of the nations" (Isa. 60:5). Chests of gold and silver coins represent a growing global market economy, and growing global economies mean new innovation and technology.

This may sound unspiritual, but the channeling of global wealth and global worship through the same doorway into the New Jerusalem does not threaten or disqualify the authenticity of the worship. We've seen Isaiah's text partially fulfilled in the famous story we recount every Christmas, of magi offering Christ their worship and their wealth.[d] Whatever predictive and spiritual interpretations these men pulled from their research of the heavens, the magi were exactly that—researchers, "devout scientists from the East"[e] specializing in "sophisticated astronomical observation."[f] As "star-gazers and wise men" they were expected to "observe and understand strange phenomena in the heavens."[g] Their work represented "the best wisdom of the Gentile world."[h] The intellectual discipline of the magi reminds us that the adoration of Christ is the proper goal of all human scientific endeavor.[i] The magi are representative, true scientists, who pursued deep discoveries in the created order and sought to walk by divine wisdom and set down a sizable pile of wealth at the feet of Christ.[j] These wealthy magi model science at its best. If the story of human innovation unfolds forever in the new creation, all of its innovators and discoverers will echo the faith and adoration of these faith-filled forefathers of true science.

While taking a more literal interpretation of the New Jerusalem, this second position takes a metaphoric interpretation of Jesus's words. We will not govern the new creation through willpower but through tools we have come to use in this world—tools we will forever be improving. In the new creation, war weapons will be recycled into farm tools.[k]

Technology Glorified

In the end, both positions have strengths. Will life in eternity pick up the technological advances of this world? Or will it be radically different and simpler? A lot of the speculations I have in my own mind will only be settled in eternity. But I lean toward the second position. The same Jesus who multiplied edible fish during his earthly ministry to feed a mass crowd stood over a charcoal grill after his resurrection to cook a few fish.[l]

We know that God's ultimate plan is "a plan for the fullness of time, to unite all things in [Christ], things in heaven and things on earth" (Eph. 1:10). Christ is "the great recapitulator," Jacques Ellul says. He will take our meager attempts at city building on earth, join them to his heavenly city, and make a true city. So as Noah midwifed tech from the first creation into humanity's restart, so too will Christ bridge tech from the cities of man into the city of God. Thus, even Ellul, a tech pessimist, favors tech continuity in the new creation when he admits that "God's plan also includes things invented by man, what he laboriously put together piece by piece, learning from experience and failure. Both his technical failures and the marvels of his cleverness. All this is 'recapitulated' in Christ, summed up in him, taken over by him. In a brilliant transfiguration all of man's work is gathered together in Christ."[m]

We know that the New Jerusalem will be gloriously material, a blend of heaven and earth. And this material existence will (I think) call for the ongoing use of tools. But I also anticipate that every material reality we think we understand in this life will be far exceeded in the superior material realities in God's city to come.[n] How technological they will be, however, is hard to say. I know for sure that we will work, for God says that his people "shall long enjoy the work of their hands" (Isa. 65:22). In the new heavens and new earth, we will enjoy the bliss of everything working together in production and mutual love. There will be no waste, no excess, no danger, and no death.

In the new creation I will have a calling. I will work, no longer under the pain of the curse, but with a joy and ease this life cannot offer. The great resistance we push against with our tools in this life will be gone, and we will work in pure freedom and delight. We will build homes and grow crops.° Will we also travel the globe in jets? Will we explore the vastness of space in rockets? I don't see why not. But even with hopes of continuity, we await greater realities that we cannot imagine now, perhaps new powers God will code into the new creation that will make all the advances of our technologies look as basic and primitive as a child's tower of LEGO bricks.

a   Rev. 21:26.

b   Matt. 17:20; 21:21–22; Mark 11:22–23.

c   Gen. 3:17–19.

d   Matt. 2:1–12.

e   H. D. M. Spence-Jones, ed., *St. Matthew, vol. 1,* Pulpit Commentary (New York: Funk & Wagnalls, 1909), 54.

f   Richard T. France, "Matthew," in *New Bible Commentary: 21st Century Edition,* ed. D. A. Carson et al. (Downers Grove, IL: InterVarsity Press, 1994), 908.

g   Craig A. Evans, *The Bible Knowledge Background Commentary: Matthew–Luke,* ed. Craig A. Evans and Craig A. Bubeck (Colorado Springs, CO: David C. Cook, 2003), 57.

h   W. D. Davies and Dale C. Allison Jr., *A Critical and Exegetical Commentary on the Gospel according to Saint Matthew, vol. 1,* International Critical Commentary (New York: T&T Clark International, 2004), 228.

i   Erasmo Leiva-Merikakis, *Fire of Mercy, Heart of the Word: Meditations on the Gospel according to Saint Matthew, Chapters 1–25* (San Francisco: Ignatius Press, 1996–2012), 1:75.

j   Leiva-Merikakis, *Fire of Mercy, Heart of the Word,* 1:105–6.

k   Isa. 2:4.

l   Mark 6:30–44; Luke 9:10–17; and John 21:4–14.

m   Jacques Ellul, *The Meaning of the City* (Eugene, OR: Wipf & Stock, 2011), 176.

n   Isa. 60:17.

o   Isa. 65:21.

6

How Should We Use
Technology Today?

SO FAR IN THE BOOK we have discovered the origin of innovators in God's plan and the origin of innovations in God's creation. We have looked at where human innovation is headed, namely, for another showdown with God. Finally, we turn to the complex ethical dilemmas of living out our Christian lives from inside the tech age.

The book of Proverbs pleads with us to find wisdom. And this wisdom is something our tech age cannot give us. Our technologies can amplify our powers, but they can't give us wisdom with those powers. As pastor Ray Ortlund says, "If we have technology but not wisdom, we will use the best communications ever invented to broadcast stupidity."[1] This is our reality. The latest iPhone doesn't upgrade our wisdom. No tech or science will do this. Wisdom is about value, and "science does not deal with questions of value."[2]

So where is wisdom to be found?

1   Raymond C. Ortlund Jr., *Proverbs: Wisdom That Works* (Wheaton, IL: Crossway, 2012), 17.
2   Yuval Noah Harari, *Homo Deus: A Brief History of Tomorrow* (New York: HarperCollins, 2017), 283.

Searching for Wisdom

Extract a few humans on a rocket-powered ark to Mars, set up a colony with nuclear generators and greenhouses, restart our species on a new planet, and we would still never answer the *why* question—why we exist in the first place. Science accomplishes incredible things, but it cannot deliver meaning or purpose. To understand why this is the case, we turn to our ninth and final biblical text.

In Job 28:1–28, we join Job's search for wisdom. Job's life has been battered, and he needs answers for his suffering. His friends have offered a lot of words but mostly hot air and vanity. Wisdom is not found in lots of words. But after a lot of words, late in the book, Job answers his friends with a vibrant soliloquy on mining techniques. Why mining? At this point in human history, mining was considered a dominant technology. When man plunged under the surface of the earth to extract minerals, silver, gold, and jewels, he flexed his dominance over creation.

Even in the ancient world, farming and ranching were modest, practical, and predictable arts. Factoring in a few variables, and assuming no drought or disaster, annual yields were fairly predictable. With stabilized grain and cattle sources in place, mankind could then invest in something calling for "random effort: irregular in routine and uncertain in result"—the practice of mining.[3] Mining is vocational gambling. The promise of extracted jackpots allure the self-driven and the ambitious speculators with dollar signs in their eyes, willing to risk life and limb to bore into any depth that promises a chance at wealth. Unearthing riches from the earth is

---

3 Lewis Mumford, *Technics and Civilization* (1934; repr., Chicago: University of Chicago Press, 2010), 66–67.

the ultimate high-risk, high-reward startup. It is a form of space exploration, not into the heavens above but into the vast unseen space under the earth's surface. It is why the miner's ambitious exploits captured the awe of his age like the first moon landing captured public awe in 1964.[4]

These ambitious ancient miners illustrate Job's desperate search for wisdom. But Job finds that wisdom cannot be found in ambitious mining techniques. That's in verses 1–11.

> [1] Surely there is a mine for silver,
> and a place for gold that they refine.
> [2] Iron is taken out of the earth,
> and copper is smelted from the ore.
> [3] Man puts an end to darkness
> and searches out to the farthest limit
> the ore in gloom and deep darkness.
> [4] He opens shafts in a valley away from where anyone lives;
> they are forgotten by travelers;
> they hang in the air, far away from mankind; they swing to
> and fro.
> [5] As for the earth, out of it comes bread [agriculture],
> but underneath it is turned up as by fire.
> [6] Its stones are the place of sapphires,
> and it has dust of gold.

Mankind is *this* ambitious. Appreciate for a moment the image here of a dusty miner inching his way down a dark shaft by a rope, swinging back and forth, squinting his eyes, holding a torch with

---

4   Bill Cotton, *Job: Will You Torment a Windblown Leaf?* (Fearn, Ross-shire, UK: Christian Focus, 2001), 118.

one hand while running the other hand across a wall to explore the dark face of rock for the first time. Farmers cultivate soils on the surface, and out of that surface comes bread. But miners explore deep caverns latent with possibilities, veins marbled and patterned within creation's rock.

The text here suggests that miners may have used water and fire to cool and heat and crack boulders apart. In either case, humans are endlessly aggressive to discover what has never been seen before. Innovations and discoveries happen in the privacy of mine shafts and back-corner laboratories, in hole-in-the-wall startups and fabled garage workbenches, and then get brought into the world.

Job's miner is a "dauntless technologist," an archetype of human ingenuity and ambition.[5] In comparison to the animal world, man sets himself apart with the desires of his mind. Animals do not pan for gold or lust after shiny jewels. Mining distinguishes humanity from all the other creatures on earth. So the keen-eyed miner plunges into realms of darkness that animals have not discovered and have no interest in discovering.

> [7] That path no bird of prey knows,
>   and the falcon's eye has not seen it.
> [8] The proud beasts have not trodden it;
>   the lion has not passed over it.
>
> [9] Man puts his hand to the flinty rock
>   and overturns mountains by the roots.
> [10] He cuts out channels in the rocks,
>   and his eye sees every precious thing.

5   Elmer B. Smick, "Job," in *The Expositor's Bible Commentary* (Grand Rapids, MI: Zondervan, 1988), 4:976.

¹¹ He dams up the streams so that they do not trickle,
and the thing that is hidden he brings out to light.

Mining is inorganic, the first entirely man-made vocational setting,
writes Lewis Mumford. "Day has been abolished and the rhythm
of nature broken: continuous day-and-night production first came
into existence here. The miner must work by artificial light even
though the sun be shining outside."⁶ Mumford calls mining the
original modern vocation, broken away from nature, distant from
animal life, hidden from the sun, and unbound from the circadian
rhythm. The mine is the first undistracted workplace, the first
windowless factory, a place sequestered for work and for work
only—a historically inhumane vocation.⁷

Mumford's overzealous criticism misses Job's celebration of ancient deep-shaft mining expeditions as one of the great human
innovations in the ancient world. Mining is ambitious due to
its risks. Few scenarios are scarier than being in a deep cavern as
waters begin to rise. But with his ambition, the miner shovels his
fear aside. Miners stop up rivers and dry out caverns in order to
tunnel downstream and explore and to bring to the sunlight gems
and wealth previously submerged. No other animal has this ambition. Only humans innovate by excavating deep into the earth
to discover new powers and riches and technological possibilities.

Humans are wired for discovery. Whether mining for gems or
colonizing on Mars, both dangerous enterprises get humans out of
bed every morning. Aspiration is fundamental to our nature. We
transcend. We explore. We invent. We penetrate nature's possibilities. The ancient miner, "by all-conquering industry and scientific

6 Mumford, *Technics and Civilization*, 69–70.
7 See George Orwell, *The Road to Wigan Pier* (New York: Mariner, 1972).

skill, he surmounts most of the difficulties in the way of his object."[8] Not by faith, human ambition alone is all we need to rip up mountains by the roots. Undaunted, we assault creation—smashing rocks, flipping mountains, redirecting rivers. Nothing stands in our way.

In the age of strip mining, this ambitious impulse may need to be held in check. But the ancient miner is a good metaphor for all human technological aspiration. Mining is the first startup. The miner is the forefather to the ambitious innovator today, and both are united by a desire to bring to light what has never been brought to light since.

In spite of this ambition, the pursuit of innovation and discovery still doesn't deliver wisdom. As Job continues, we see that wisdom cannot be discovered from *within* creation itself.

[12] But where shall wisdom be found?
    And where is the place of understanding?
[13] Man does not know its worth,
    and it is not found in the land of the living.
[14] The deep says, "It is not in me,"
    and the sea says, "It is not with me."

Man wields godlike technological powers over creation. We "force nature to disclose herself" so that we can "discover her secrets."[9] Nature's secrets include new scientific discoveries, new material wealth, and new powers to enable new tech. We can upend a mountain and peek underneath, but we won't find divine wisdom. We can

---

8   A. R. Fausset, *A Commentary, Critical, Experimental, and Practical, on the Old and New Testaments* (London: William Collins, Sons, n.d.), 3:68.

9   Herman Bavinck, *The Wonderful Works of God* (Glenside, PA: Westminster Seminary Press, 2019), 16.

dig deep shafts into the earth, and we may find gold or silver or bronze or iron or precious jewels, but we won't find the purpose of life. Even if we become billionaires by the gold and jewels we unearth, wisdom is not for sale.

> ¹⁵ It cannot be bought for gold,
>   and silver cannot be weighed as its price.
> ¹⁶ It cannot be valued in the gold of Ophir,
>   in precious onyx or sapphire.
> ¹⁷ Gold and glass cannot equal it,
>   nor can it be exchanged for jewels of fine gold.
> ¹⁸ No mention shall be made of coral or of crystal;
>   the price of wisdom is above pearls.
> ¹⁹ The topaz of Ethiopia cannot equal it,
>   nor can it be valued in pure gold.

The entire material universe is made from nothing. This means that you can flip every mountain and dig into the earth and unearth billions of dollars of its wealth for yourself, but you will never find creation's true value. Made from nothing, the contingent creation itself cannot answer for its own purpose, meaning, or reason for existence. Our scientists can study the activity of tiny particles, measure the vast space of the cosmos, or discover new power sources, but they will face the exact same riddle as the ancient miner. The new discoveries of the Large Hadron Collider, the latest images from the Hubble Space Telescope, and the collected artifacts of the ancient miner all tell us a lot about the material world. But none of them can explain creation's meaning or reason for being. Neither can bars of gold, coins of silver, and bags of jewels buy these answers.

So if wisdom cannot be purchased with wealth or located with a metal detector, then where do we discover it? That's the next question.

> [20] From where, then, does wisdom come?
>   And where is the place of understanding?
> [21] It is hidden from the eyes of all living
>   and concealed from the birds of the air.
> [22] Abaddon and Death say,
>   'We have heard a rumor of it with our ears.
>
> [23] God understands the way to it,
>   and he knows its place.
> [24] For he looks to the ends of the earth
>   and sees everything under the heavens.
> [25] When he gave to the wind its weight
>   and apportioned the waters by measure,
> [26] when he made a decree for the rain
>   and a way for the lightning of the thunder,
> [27] then he saw it and declared it;
>   he established it, and searched it out.

Wisdom is found in God because he is the sovereign Creator. He patterned creation within himself. He made all things from nothing. Wisdom predates mountains and oceans and farmlands and the entire material universe.[10] Wisdom is eternally present within the eternal self-delight of God. Like the ancient miner, in our search for wisdom we soon discover that the meaning of creation

---

10 Prov. 8:22–31.

is not found in creation. The meaning of creation is found in the Creator.[11]

This text in Job is another reminder that we are witnessing God's creative brilliance when we notice how he designed wind and oceans and lightning and thunder and rainfall. He draws the boundary lines around oceans and seas and lakes.[12] Of course a hurricane swell can push water 10 miles inland. God allows and permits occasional overreach. But on an average day you know where to find the beach. To these shores God patterned *prevailing winds* and *trade winds*, and by implication installed the first cause of intercontinental navigation. To these gales God patterned *lightning*, and by implication installed the first cause of our electrified cities.

A Digression

If you'll allow a brief digression, these same patterns in the *wind* and *lightning* extend our inventive imitation beyond Job's list. They connect to our innovations in solar power, wind turbines, hydroelectric power plants, and hydrogen fuel cells. We've looked at several already, but here are a few examples of how God's patterns shape our innovations.

First, think of *navigation*. For millennia navigators steered ships by fixed points in the heavens. We imitate this pattern by GPS satellites in the heavens that send out signals to guide the navigation apps on our phones.

Now think of *thinking*. The human brain crackles electronically with eighty billion neurological links that form our conscious (and subconscious) awareness. This same pattern is reflected in the electrical power surging through your smartphone's brain, its

11 Rom. 1:18–23.
12 Gen. 1:9–10; Prov. 8:29.

computer chip. And our brain processing has inspired a new line of neuromorphic chips, futuristic processors that will more closely resemble the neurons and synapses of our brains.

And now think of *electricity production*. God patterned *nuclear* forces that we use today. I live in the nuclear-powered city of Phoenix, a city that Ellul would call one of America's atomic cities, artificially erected in the desert.[13] Artificial or not, uranium was excavated and enriched, and now atoms are splitting and my computer works. This process may sound fancy, but it's the junior-varsity version of what is possible.

My computer runs off electricity generated by *nuclear fission*, the splitting of atoms. But a more potent possibility could come in the future—*nuclear fusion*, the merging of atoms. Fusion yields more power and leaves less waste. But its development is slow because the forces at play are harder to control and the materials we need are scarcer. So where did this wildly ambitious idea for nuclear fusion come from? It came from the sun. Nuclear fusion is how God keeps the sun ignited and all the larger stars in the known galaxies. God said, let there be nuclear fusion and there was nuclear fusion, a power as old as light. We are learning from his pattern. Nuclear power is not a man-concocted artifice.

Thanks to the pattern of the sun, nuclear fusion may power our cities in the future, if we can excavate enough helium-3, a rare element on earth but abundant on the surface of the moon. Moon mineral mining may be the secret to our future power. The sun has inspired us, and perhaps automated robots controlled from the earth will excavate minerals on the moon's surface for us to electrify our cities more cleanly in the centuries ahead. If we pull

13 Jacques Ellul, *The Meaning of the City* (Eugene, OR: Wipf & Stock, 2011), 155.

all this off, who gets the glory? Not the one who discovered nuclear fusion, but the one who patterned its reality into being. The glory will go to the one who made the sun and wind and lightning and the seas, to the one who made all things.

But don't get your hopes up on all this talk of electricity, because our electric tools that seem so cutting-edge today will one day look as primitive to us as steam engines. Think of *data storage*. We know that God, with infinite care, knit each of us together in our mother's womb.[14] This weaving metaphor makes sense to primitive cultures who wove rugs, not to the technopolis that builds robots with robots. But today we can say that God was programming us in our mother's womb, coding 1.5 gigabytes of information into each of most of the cells in our body. All of this DNA data combined tallies up to about 150 zettabytes of storage in the average human.[15] This means that the data storage capacity inside my cells alone could hold the collected man-made digital data in the universe—every movie, video, photo, database, book, magazine, web page, and every 1 and 0 of digital code. We've been using electronic computer storage for data for decades. But DNA biological storage in bacteria cells may be the future of mass data storage.[16] And who invented biological data storage?

God is the meaning of the universe. We must see him and see his patterns. We must train our eyes to see creational patterns like these because how we relate to the natural world reveals how we relate to the Creator. Our worldviews break into two broad categories of faith and unbelief—the *mimetic* and the *poietic*. The *mimetic*

14 Ps. 139:13.

15 According to the musings of biologist Yevgeniy Grigoryev, "How Much Information Is Stored in the Human Genome?" bitesizebio.com (Mar. 16, 2012).

16 Sang Yup Lee, "DNA Data Storage Is Closer Than You Think," scientificamerican.com (July 1, 2019).

worldview sees a world of inherent meaning and preexisting patterns of reality to be acknowledged, respected, and followed by man. We imitate and conform to meaningful patterns beyond us. On the other hand, the *poietic* worldview sees the world as raw material for each person to make of himself whatever he wishes. In this second case we manufacture meaning for ourselves and live from a self-centric worldview in which "transcendent purpose collapses into the immanent and in which given purpose collapses into any purpose I choose to create or decide for myself." What it means to be human is rendered down to merely "something individuals or societies invent for themselves."[17]

Two very different tech trajectories are represented in these worldviews. The Übermensch is willfully blind to God in the created patterns. But the Christian has eyes to see the Creator's patterns, to acknowledge that the ultimate meaning of the material creation is God, and to know that our making must always submit to the Creator's realities. Wise living in the tech age calls us to see and appreciate God's patterns in his creation as we make our attempts at imitation.

Back to Job

God creates and governs his creation in wisdom and wealth, and new technological possibilities can be wooed from creation by humans. But the wisdom we need to thrive is not wooed from creation by humans. We can harness many of the incredible powers of creation through our know-how, and we can travel into the earth, and we can travel to the moon. But the discovery of wisdom requires us to accomplish something far beyond technical mastery.

---

17 Carl Trueman, *The Rise and Triumph of the Modern Self* (Wheaton, IL: Crossway, 2020), 39–42.

This is the leading dilemma of Job 28 and the leading dilemma of our technological age. The wisdom we need is beyond the reach of our pickaxes and our diamond-tipped drill bits.

Apart from God and without wisdom, you can be a tech-savvy fool. Even fools can land windfalls and sell startups for millions of dollars. But ambition and wealth and technical mastery must never be confused with wisdom. You can set out today to exceed the invention, wealth, and power of Elon Musk, and you can remain a fool without wisdom, a lost soul who cannot make sense of the meaning of life. Why? Because you turned to innovation and possibly wealth and power to find the meaning of the universe and the purpose for your life. And you didn't find it. You thought the power to unearth diamonds would deliver an answer to the meaning of life. It won't.

So where can we find this wisdom?

28 And he said to man,
"Behold, the fear of the Lord, that is wisdom,
    and to turn away from evil is understanding."

If we put the miner (of Job 28) alongside the farmer (of Isaiah 28), we discover something essential. The process of pulling riches from the earth or coaxing a harvest from the land was coded into creation by the wisdom of God. But becoming an expert miner or farmer doesn't make one wise. Wisdom is found in the Creator, not in pilfering his creation. We must behold the Creator behind all our innovative endeavors. We can take God's created patterns and convert them into power and wealth. But wisdom is found in God.

At the end of the search, here is what Job discovered. The secret to finding wisdom is in God's reality, in his weight and glory. As we fear God, he becomes the dominant reality in our lives, the orbital center of who we are and what we live for.

When God becomes the gravitational mass at the center of your life's orbit, you will discover the right place for the planets of science and tech. *He* is the end of all the created gifts we have been given. The opposite of fearing God is replacing God. As we discovered earlier in Jeremiah 2:13, God says that "my people have committed two evils: they have forsaken me, the fountain of living waters, and hewed out cisterns for themselves, broken cisterns that can hold no water." The greatest sin in the universe is to turn away from God and ignore him for some vanity engineered in the mind of man. In contrast, the fear of God is wisdom. In him our search ends.

Two Revelations

My earlier book on smartphone overuse served as a warning but also as an optimistic vision for the long-term value of digital media. The emphasis was intentional, and it's not only about smartphones. In my broadest understanding, all our technological advances include three stages: (1) *discovery*, (2) *production*, and (3) *adoption* of those new powers into our lives to amplify our native dexterity. Such a definition of tech is relatively common. Missing is the fourth stage that must be added: (4) *adapting* these new powers to human flourishing. In a fallen world, our adaptation will always lag behind discovery, production, and adoption. The smartphone vibrantly illustrates this. We are living through stage four now, learning to optimize the iPhone into our lives—that is, narrowing the perimeter of its uses to its true usefulness—a slow lesson we are still figuring out. We eventually will. But not without struggle. This is the paradigm we face with every new innovation.

But without wisdom, we are left in the dark as to whether our inventions are truly helping or hurting human flourishing. We try to invent but not overinvent. We try to flourish but also not create too many Frankenstein monsters along the way. We want to hear the Creator's voice and correct when we overreach, because if we go beyond his voice we will end up polluting the world, maiming others, and accidentally killing ourselves. Some human harm is inevitable in innovation. So we must stay attuned to the fallout of every technology, both physically and spiritually, both through general revelation and special revelation.

God patterned the material universe, and he continues to create innovators. The Creator links the natural order and innovation. With the Creator in place, our innovations are held in check by the dialogue with creation. So a Christian approach to technology includes a close dialogue with creation itself. By listening to creation, we can discover new possibilities while also keeping creation care in mind. We must listen to creation for our cues of possibility and to hear the limits of what we should and shouldn't do. When we discover that a useful chemical causes skin cancer, we replace it. When nitrogen leeches into groundwater and kills off marine life, we rethink it. When asbestos is found to cause lung cancer, we remove it. And when freon pokes holes in the ozone, we ban it. But all these corrections operate at the level of general revelation. You don't need to know the wisdom of God or the meaning of life to make these adjustments.

Christians bring another revelation into the story of technological advance. We bring wisdom. Believers hear general revelation (in the earth and heavens), *and* we hear special revelation (in Scripture). As soon as the searchlight of divine revelation stops shining into any industry, whether that's commercial farming or video game development, that industry will deteriorate and operate by lax moral standards. That industry or corporation will be governed by greed. It will dehumanize eternal souls.

There are a million ways to use innovation, but technology is used best when we follow the lead from creation and restore what is broken. The world is not the product of evolutionary chance; it's the product of the Creator's intentional design. We can honor that system when technologies are used to fix what appears to be broken within the created order. A whole book of application is needed to trace out this ethic in genetics, reproduction, chronic diseases, and previously "irreversible" injuries. Technology puts into our hands new powers to break creational patterns and accomplish what is unnatural. But it also grants us new powers to restore the normal creational patterns we discover by general revelation.

Hearing the Creator speak through creation is how we discover new possibilities. Fearing the Creator, listening to him through the life of his Son and in the revelation of his word, is how we refine our practices to serve the flourishing of humanity *and* creation. Both voices—the creation and the Creator—must be heard. We must listen in stereo, with both earbuds, for new possibilities and for governing limiters—not as some isolated spiritual exercise but as an act of love to protect the health of all civilization. General revelation (God's voice through creation) and special revelation (God's voice in his word and Son) work in tandem to preserve the human race, "first by sustaining it, and . . . second by redeeming it," and together by serving the ultimate end of glorifying God in all his revealed beauty.[18]

18 Bavinck, *Wonderful Works of God*, 22.

Ethics, Technology, and Wisdom

With or without wisdom, man's aspiration for new technology is unquenchable. In the last century, humanity has found new ways to largely end hunger, infection, and war. So what's next on the docket of human aspiration? In the words of Yuval Noah Harari:

> Success breeds ambition, and our recent achievements are now pushing humankind to set itself even more daring goals. Having secured unprecedented levels of prosperity, health and harmony, and given our past record and our current values, humanity's next targets are likely to be immortality, happiness, and divinity. Having reduced mortality from starvation, disease, and violence, we will now aim to overcome old age and even death itself. Having saved people from abject misery, we will now aim to make them positively happy. And having raised humanity above the beastly level of survival struggles, we will now aim to upgrade humans into gods, and turn *Homo sapiens* into *Homo deus*.[19]

Having largely ended starvation, disease, and global warfare, we will turn our attention to anti-aging, chemically induced happiness therapies, and brain and body augmentation until we are no longer *human beings* but *human gods*. Not like the omniscient God but like Greek gods, superhuman beings. As this happens, as man self-evolves and self-creates, and as the moral dilemmas complexify, the Bible will finally become irrelevant, says Harari.

What will happen to the job market once artificial intelligence outperforms humans in most cognitive tasks? What will be the

19   Harari, *Homo Deus*, 21.

political impact of a massive new class of economically useless people? What will happen to relationships, families, and pension funds when nanotechnology and regenerative medicine turn eighty into the new fifty? What will happen to human society when biotechnology enables us to have designer babies, and to open unprecedented gaps between rich and poor? You will not find the answers to any of these questions in the Qur'an or sharia law, nor in the Bible or in the Confucian *Analects*, because nobody in the medieval Middle East or in ancient China knew much about computers, genetics, or nanotechnology.[20]

Before diving into very specific ethical categories, we must grasp three important principles.

First, Harari is right that we should prepare for some mind-bending ethical dilemmas. The God who makes sure that the ravager has a sword has also patterned a creation that can produce far beyond what Christians will endorse morally.

The reality is that our ethics will never catch up to our techno-possibilities. God saw this problem from the start. Adam and Eve entered the world naked and unashamed.[21] They were childlike. So when God banned the first couple from eating from the tree of the knowledge of good and evil, he was not forever limiting man's scientific discovery (per Sagan), but temporarily holding back the scientific discovery of creation until man had matured into an adulthood capable of managing all of its potent possibilities. The tree was meant for food, but not immediately. Man in his infancy was ill-prepared for science, technology, and all the ethical dilemmas they would usher into the world. This awakening would come

20 Harari, *Homo Deus*, 271.
21 Gen. 2:25.

later, as man matured. For now, the couple would be happy in their childlike obedience to God in the garden. But like a child disobeying his father, Adam and Eve reached for a greater knowledge of the world that they were ill-prepared to handle ethically.[22] Thus, traced back to the very first sin, our technological ethics have never kept up with our new technological possibilities—and they never will, a haunting tragedy for sinners living in such a potent world. We especially feel the pressure now because "the world-altering powers that technology has delivered into our hands now require a degree of consideration and foresight that has never before been asked of us."[23] Life in a fallen world means that our ethics will never catch up to tech's latest possibilities.[24]

Second, new technologies don't generate new questions about those technologies. In fact, this is one of the great concerns with technology. Right now, I can buy a DIY gene engineering CRISPR kit online for less than $200 and have it mailed to my house. I can teach myself how to manipulate human DNA. And the kit requires no ethical answers. Ethical boundaries are not packaged with new technologies. Technological "progress" rarely slows down for unanswered ethical questions.

Christian ethicist Oliver O'Donovan explains this dynamic when he writes, "If a moral 'issue' has arisen about [a] new technique, it has arisen not because of questions the technique has put to us, but of questions which we have put to the technique."[25] In other

---

22 Umberto Cassuto, *A Commentary on the Book of Genesis (Part 1): From Adam to Noah*, trans. Israel Abrahams (Jerusalem: Magnes Press, 1998), 112–14.
23 Carl Sagan, *Pale Blue Dot: A Vision of the Human Future in Space* (New York: Ballantine, 1997), 317.
24 Thanks to Alastair Roberts for prompting this paragraph.
25 Oliver O'Donovan, *Resurrection and Moral Order: An Outline for Evangelical Ethics* (Grand Rapids, MI: Eerdmans, 1994), 93.

words, society often embraces new innovations while wearing ethical blinders, galloping straight ahead without asking what path the technology might be on in the first place. Technologists whip the inventions to go faster, but Christians stop to examine the risks. We critique tech because tech does not self-critique.

Third, and most importantly, Harari's dismissal of the Bible is terribly premature. As O'Donovan points out, innovations don't actually raise new questions; they call for new clarity about old priorities. This is a critical difference. The world is based on old, precious, fundamental truths of life found in God's word, which stands forever.[26]

So we can imagine a high-tech, Wild West fantasyland where any paying patron can indulge in the fantasy of raping or killing lifelike robots. Such a fantasyland says something about the advance of robotics. But it says a lot more about the evil violence inside the human heart. A 90-pound sexbot with customizable labia and breasts, designed with the overall shape and sounds and behaviors patterned after a surgically enhanced porn star, doesn't simply raise questions about whether or not intercourse with a machine is holy. It raises questions over what sex is for, where it flourishes, and how its misuse damages souls.

Harari mistakenly assumes that the ancient Bible is ethically relevant for the digital age only to the degree that it speaks the lingo of the tech age. Rather, the essence of the "new dilemmas" of our innovations forces us to find new clarity on what it means to be an enfleshed human being. It's not hard to make a list of other relevant questions.

- Where do we find happiness?
- What value is found in the material body, even a broken one?

26 Isa. 40:8.

- What does it look like for souls to flourish in a material-centric culture?
- What does it look like to care for the health of our bodies?
- What does it look like to love the poor and not exploit them?
- What role does vocational purpose play in human flourishing?
- What does it mean to be married?
- What does it mean to be a parent?
- What does it mean to love others?
- What does it mean for a fetus to be a person at conception?
- What does it mean to be a woman and not a man?
- What does it mean to kill an enemy in a war?
- Why do we preserve personal privacy?
- Why do we preserve religious liberty?
- What does it look like to pursue justice for our neighbor?

These are just a few questions that will be raised over and over again by technology. New chemicals, atomic powers, AI, weaponized drones, and automated robots—new tech simply moves ancient truths about human flourishing back into the foreground of our tech-age ethics. As we seek wisdom and meaning and purpose, the eternal relevance of God's wisdom in Scripture will shine in the tech age.

How Then Shall We Live?

Our rapidly accelerating technological powers to modify creation (and now ourselves) call for several ethical convictions. With open Bibles, here are fourteen to consider.

1. We respect the gifts of science and non-Christian innovators.
Many scientists are hostile to Christianity, but not all of them.

The Human Genome Project was a fifteen-year endeavor that completely mapped and sequenced three billion base pairs of DNA. It stands as one of the most important scientific and medical projects of our lifetime. We have only begun to understand human genetic coding and what we can detect and change and fix. The project, completed in 2003, was headed by Francis Collins. He and his team mapped out all three billion letters of the human genome with a sequencing that "will be seen a thousand years from now as one of the major achievements of humankind."[27] Collins called it "a stunning scientific achievement and an occasion of worship."[28]

Collins is a Christian. Science may attract the godless, but science also drove Collins from atheism into the arms of God. Science could never answer his ultimate questions about the universe. Science gave him knowledge without wisdom. Collins, an agnostic, then atheist, then Christian, came to see faith in God as "more rational than disbelief."[29] He's now one of the world's most celebrated geneticists and brain scientists. "People said my head was going to explode," he said, looking back to his conversion, "that it would not be possible to both study genetics and read the Bible. I've never found a problem with this at all, despite the way in which some scientists have caricatured faith to make it seem incompatible." Long before the Enlightenment, hostile tension simmered between faith and science. Scientists often negated faith, and the church often rejected scientific discovery. So how do we proceed today in the age of innovation? "I don't want to see a future where this science-versus-faith conflict leads to a winner and a loser," Collins said in an interview. "If science

27 Francis S. Collins, *The Language of God: A Scientist Presents Evidence for Belief* (New York: Free Press, 2007), 122.
28 Collins, *The Language of God*, 3.
29 Collins, *The Language of God*, 30.

wins and faith loses, we end up with a purely technological society that has lost its moorings and foundation for morality. I think that could be a very harsh and potentially violent outcome. But I don't want to see a society either where the argument that science is not to be trusted because it doesn't agree with somebody's interpretation of a Bible verse wins out. That forces us back into a circumstance where many of the gifts that God has given us through intellectual curiosity and the tools of science have to be put away."[30]

When the Bible and science meet an impasse, we are often told that science is to be declared the winner. This conclusion sounds too similar to what I hear from atheists.[31] Rather, I think we should critique applied science with open Bibles. But Collins is right; we aim for mutual respect. The Christian honors the scientist's discoveries. The scientist honors the Christian's ethical concerns.

In all science and innovation we can discern what is good, because in what is good, we can sense a gift from the Spirit, coming to us through men and women who may spiritually remind us of Cain and his heirs.

2. We expect human innovation to serve an ecosystem.

As Psalm 104 shows, God created and now sustains a whole ecosystem. He patterned it all. He made the moon and darkness, the sun and daylight, and the seasons. He feeds the animals and leads mankind to work from morning until evening.[32] He created the raw materials for man to make wine, oil, and bread.[33] All of man's

---

30 Jebediah Reed, "A Long Talk with Anthony Fauci's Boss about the Pandemic, Vaccines, and Faith," nymag.com (July 1, 2020).

31 See, e.g., Carl Sagan, *The Demon-Haunted World: Science as a Candle in the Dark* (New York: Ballantine, 1997), 277–79.

32 Ps. 104:19–23.

33 Ps. 104:14–15.

work is the work of God.[34] God gave us ships to zip over the sea just as he created Leviathan to play within the sea, and both have their place.[35] Stand back and look at this entire system and praise God! Mankind's labor and science and innovation must always fit within a larger ecosystem.

Technology originates from the created order. But the rules and governors for technology are not in the technologies; they reside within the created order. Our innovations must exist within an ecological dialogue with God as we properly care for and cultivate creation.[36]

At its best, the most revolutionary technologies display the glory of God. So when Kevin Kelly suggested that "we can see more of God in a cell phone than in a tree frog," he was right in one sense.[37] The smartphone certainly reflects the Creator's glory. But it reflects God in a more basic sense than the tree frog. The tree frog was created by God, and it adapts to the Creator's biological patterns in creation, and it does so in a very refined sense. A tree frog lives entirely within the created order and consistently conforms to it. The tree frog lives obediently to the Creator's voice. But as smartphone users have discovered, the iPhone introduces a host of physiological problems like sleep deprivation, increased anxiety, soaring depression, personal alienation, eye strain, shallow breathing, and bad posture. (And the iPhone is changing us in a dozen spiritual ways too.[38]) We'll get all this phone misuse figured out eventually and advance to the socially healthy behaviors of a tree frog. But for now, the smartphone, and our overuse of it, will

34 Ps. 104:23.
35 Ps. 104:26.
36 Gen. 1:28–31.
37 Kevin Kelly, *What Technology Wants* (New York: Penguin, 2011), 358.
38 See Tony Reinke, *12 Ways Your Phone Is Changing You* (Wheaton, IL: Crossway, 2017).

often fall short of conforming to our biological flourishing. It will fall short of the glory of God.

At root, human sin is the destruction of the natural order.[39] And grace sets out to restore nature.[40] There's an unmistakable connection between human rebellion and the resistance of the created order—in favor of trends that are self-destructive of our biological lives. Regeneration begins to restore us from the inside and teaches us to tune our hearts to the Creator's voice on the outside. Human flourishing calls for opposition to sin, opposition to what breaks down the natural structures of biological flourishing. Christians are given this role in society because our enemy is not nature but sin's distortion of nature. Grace restores nature by confronting the human sin that distorts nature. As God's grace teaches us, and as his Spirit convicts us, we will resist innovations that fail to honor creational patterns. But our discernment will always be catching up to the tree frogs.

3. We expect to witness technological overreach and commit to correcting as we go.

Near the end of the hotly contested presidential election of 2020, theologian Wayne Grudem was invited to instruct a church to participate in the voting process, making the case for reelecting Donald Trump. There he rhetorically asked the audience about fossil fuels, a key theme in the election. "Do you think that God put these amazing and accessible energy sources—coal, oil, and natural gas—in the earth so that we could use them, but that he booby-trapped them so that they would destroy the earth?" The crowd laughed and clapped in agreement with his premise. "I don't

---

39 Rom. 1:18–32.
40 Herman Bavinck's major theological motif.

think so."[41] Now, the first half of the statement is gloriously true. Fossil fuels are intentionally gifted to us by God, and they are amazing and accessible and have changed our lives in countless ways. (Have you ever stopped to worship God for the gift of fossil fuels?)

But the second half of the question is a little more problematic, and you can see why simply by replacing "coal, oil, and natural gas" with "uranium." Switch the categories and the answer changes. In a number of scenarios, humanity could extinguish itself through a global thermonuclear war and its consequences. So the answer to the rhetorical question is not as simple as it sounds, because natural accessibility does not equate to foolproof safety. Our most potent discoveries call for self-limitation.

Thinking back to the ancient farmer of Isaiah 28, God's lessons were "expressed in all the laws of nature, in the character of air and of soil, of time and place, of grain and corn." But as the farmer learned these lessons, he was still "liable to mistake and error."[42] And when the ancient farmer made mistakes, he could hurt his crop and damage the lives of those around him through starvation. But the impact of mistakes and errors today are multiplied by the size of the industrial complex. Our technological mistakes today hurt, maim, and kill at scale. The stakes for misreading the Creator are higher than ever. Therefore, ecological awareness must be on the church's radar.

Theologian J. I. Packer point-blank called the Western world a "technological monster, raping the planet for financial profit and generating horrendous ecological prospects for our grandchildren."[43] Power can be overused, and creation can be overworked. The Cre-

41 Jack Hibbs, "Answering a Friend's Objections to Voting for Trump, Dr. Wayne Grudem," *Real Life with Jack Hibbs*, youtube.com (Oct. 18, 2020).

42 Bavinck, *Wonderful Works of God*, 49–50.

43 J. I. Packer, "Our Lifeline," *Christianity Today*, October 28, 1996, 23.

ator teaches the farmer to plow once, not all the time. Overwork the soil and you can unleash on yourself a catastrophic dust bowl. Good farming is about intuition. Farming is a dynamic dance in tune with creation's seasonal ebbs and flows. Even with all the shiny technology, good farming rides on instinct. Technology requires ecological discretion.

Because all human activity changes the balance of creation, we will overreach with our innovations and need correction. But we will never outreach God's providence. Packer warns that "if, for instance, we continue in industrial life in a way that actually does produce major global warming in the way that they're warning us it might, well, that will mean certainly more storms, more natural disasters, more violence in the natural order. And there will come a time when we shall have to whisper to ourselves, 'We brought it on ourselves.' And in that sense, we must accept responsibility for it. But it's still under God's sovereignty."[44] Yes. Not one sparrow falls dead from the sky apart from God's will and timing.[45] But if we find a pile of sparrows under a wind turbine, we should launch an environmental inquiry.

As innovations remove us from the land, our ears must be even more carefully tuned to the Creator's voice in his creation. Future innovations will make this possible in ways hard to imagine now. Large-scale farming has automated in many ways, with GPS-navigated combines and precision-guided planters. We tend to think of farming in macro terms, big tractors doing bigger things faster, fertilizing and harvesting in bulk. But AI robots may eventually steward the land better than we can. Future machines will granularize farming. On the horizon are robots with more tactile skills, like intelligent

44 "John Piper Interviews J. I. Packer," desiringGod.org (July 28, 2020).
45 Matt. 10:29.

harvesting robots, AI-equipped brains with eyes and arms and digital discernment to know a ripe fruit to be picked today from an unripe fruit to be picked later. Small, automated robots could inch their way through fields, pulling weeds one by one. This would cut down on pesticides. Perhaps the same robot could name each plant too, stewarding each of them, documenting into a database the plant's particular growth and health and needs. Crop care at this granular level would impress Tom Bombadil—and this tech could easily get adopted by organic farmers too. Imagine the water and fertilizer and chemical conservation that AI robotic automation could bring.[46]

Nature is worth such care. Creation is more than raw materials for our manipulation. Nature is a living organism we don't fully understand, and much of its advanced chronobiology remains a mystery. Earth's plant and animal and insect worlds are biological networks of life and communication that we have hardly come to appreciate. Maybe we will rely on a swarm of nano-satellites in low earth orbit to help us monitor in real time the earth's "planetary hygiene."[47] Perhaps we can invent new ways of hearing the Creator, as he continues to speak to us through the catechism of his creation. As we disrupt the balance of creation, we will need a conscious capitalism and creative ways to simultaneously "save the economy from freezing and the ecology from boiling."[48] In other words, we will learn to correct as we go.

4. We expect technological progress to honor God's design for the body.

The Jetsons leased a domestic robot maid. Her name was Rosey, a blue, well-worn generalist robot programmed to cook, clean, and

---

46 Kevin Kelly, "The Future of Agriculture," youtube.com (Aug. 26, 2020).
47 Sagan, *Pale Blue Dot*, 71.
48 Harari, *Homo Deus*, 214.

parent when necessary. She was a cartoon vision from 1962 of what 2062 could look like. But generalist robots like Rosey are not likely to do our laundry and our dishes and our cooking anytime soon. "There is some fantasy that we can make an AI robot that is superior to humans in all dimensions. That's just technically, engineeringly impossible," says Kelly. "We cannot make a machine that excels humans in all dimensions. You can make a machine that can run faster than the human. You can make a machine that can jump higher. You could make a machine that can crawl lower. But you can't make a machine that does all those things at the same time because there's an engineering trade-off. In addition to us being fairly powerful, we're incredibly flexible."[49]

After three years of experiments on the surface of Mars by the Curiosity rover, one of its engineers estimated that the same workload could have been accomplished by one human in about a week.[50] Compared to robots, the human body is powerful, flexible, and efficient. It can run for sixteen hours every day, optimizing its mere quarter horsepower of energy, governed by a brain that requires less power than a light bulb. We are tremendously efficient, adaptable, and powerful. Robots are not. "We don't know how to make a flexible, powerful machine," says Kelly. "Those two things are normally not what we optimize. There's no reason to try to make a machine totally like us—flexible, strong, fast, long duration, low-powered—because we can make more of us very easily. Most robots will be different from us in many, many ways. They will be better than us in certain narrow ways."[51] But they will remain alien to us.

49 Kevin Kelly, "The Future of Robots," July 8, 2020, youtube.com.
50 Bobak Ferdowsi in Netflix's *The Mars Generation*, 2017, directed by Michael Barnett, produced by Austin Francalancia and Clare Tucker.
51 Kelly, "The Future of Robots."

The human body is a remarkable design. And we will be given more power to optimize this body for sure. The age of anabolic steroids may give way to an age of genetically modified superhumans. As we learn how to biohack the human body and optimize it for strength and speed, science is beginning to disrupt athletics, a venue where human competition has long assumed biological and chemical neutrality. What happens to sports when kinetic modifications are genetically rearranged? What happens when rapid strength and speed can be gained through genetic changes?[52] Will genetic modifications eventually bring sports to a gradual end? Or will they, like the steroid era in baseball, make a sport more popular and exciting? Will athletes become the first genetically enhanced class of superbeings?

In response to debilitating challenges, neurological implants may help remedy strokes, brain seizures, blindness, deafness, and even paralysis. They may even help alleviate depression, anxiety, and addictions. But ethics get sketchier when we talk about the use of these implants to augment healthy people. Augmentation does not seem to trouble inventors like Elon Musk, who is looking to accelerate human communication. It requires a lot of work to compress ideas accurately for others in concrete words, and the process of expression operates at a very low data rate. So what's Musk's proposed fix? A brain implant to allow for uncompressed impressions and concepts and thoughts to pass from one brain to another through "non-linguistic consensual telepathy."[53]

Consensual telepathy through brain-machine interfacing is a "heady" sci-fi enhancement plagiarized from the pages of Dr.

---

52 See Se-Jin Lee's discovery of myostatin.

53 "WATCH: Elon Musk's Neuralink Presentation," CNET Highlights, youtube.com (Aug. 28, 2020), 1:04:03–. I should note that the Spirit communicates through some sort of inexpressible intercession "too deep for words"—a whole wordless (ἀλάλητος) channel we cannot comprehend (Rom. 8:26).

Filostrato. But this is the trajectory of technology—making our innate powers superhuman. "Twentieth-century medicine aimed to heal the sick. Twenty-first-century medicine is increasingly aiming to upgrade the healthy."[54] Such biological enhancements are unnatural, says Kevin Vanhoozer. "To heal is to intervene therapeutically by aiming to correct a biological or biochemical defect. In contrast, to enhance is to improve normal function, to go beyond the natural."[55] There's a delicate balance to be maintained in preserving and restoring natural functions—and grave consequences to enhancing the natural into the super-natural. Supernatural enhancements bear supernatural consequences. For the "enhancement of the body is the disenchantment of the soul."[56]

We may eventually modify our bodies in ways that are helpful and necessary. But we will never reach a point where the human body is a disposable machine. Nazi eugenics proved wicked in their attempts to isolate a superior race, but that noxious spirit can re-emerge in the tech age. Today, prenatal screening is used on preborn children to identify possible genetic defects, using a science-based rationale that leads to the slaughter of countless children who are predicted to have Down syndrome, precious people who are notoriously the happiest populace on earth.[57] Genocidal slaughter hides behind the veil of scientific objectivity. Every body, broken as it may be in this fallen world, is a fully valuable human being, reflecting its Creator. Technologies that devalue the human body will never honor the Creator.

54 Harari, *Homo Deus*, 353.
55 Kevin J. Vanhoozer, *Pictures at a Theological Exhibition: Scenes of the Church's Worship*, Witness and Wisdom (Downers Grove, IL: IVP Academic, 2016), 256.
56 Vanhoozer, *Pictures at a Theological Exhibition*, 260.
57 John Knight, "The Happiest People in the World," desiringGod.org (Mar. 20, 2015).

5. We commit to our chief vocations of love.

A long time ago, Herman Bavinck came up with his own list of perennial questions humans will always need to answer, particularly these six:

- What is the relationship between *thinking* and *being*?
- What is the relationship between *being* and *becoming*?
- What is the relationship between *becoming* and *acting*?
- Who am I?
- What is the world?
- What is my place and task within this world?[58]

We return to these same questions in the age of innovation, because innovations never answer these questions. But the answers to these questions also require the wisdom that determines the purpose of our lives. And this wisdom is crucial when we can no longer turn to the market to identify our marketable skills, where we so often look for our identity and purpose.

When it comes to understanding our place in the world, a major change is brewing for all of us. AI is changing more than facial recognition and national defense strategies. It is changing everything. Computers are on the verge of self-training and self-improving. AI learns by studying us. It studies you whenever you speak to a personal assistant like Alexa or Siri. It watches and learns your behaviors and preferences when you type simple Google searches. AI watches your Netflix habits to gather your personal interests, inform a database, and guess what media you might like

---

58 Herman Bavinck, quoted in John Bolt, *Bavinck on the Christian Life: Following Jesus in Faithful Service*, Theologians on the Christian Life (Wheaton, IL: Crossway, 2015), 122–23; formatting added.

served up next. Amazon AI seeks to sell you new things based on who it thinks you are. AI shields you too—spotting spam to keep it from overtaking your inbox. AI computers mastered chess and have moved on to mastering video games, learning how to win by creating new tactics beyond those of human players. But AI is not all fun and games and consumables.

As wearable tech becomes ubiquitous, everything we do and touch and own is connected to a digital neurology, a funnel flow of data. All our behaviors are being digitized. You, your wearables, your behaviors, your life decisions—they are all feeding new data streams. Our collective data flow syncs to a database of all other human behaviors. A collective database of human conduct is making targeted marketing nearly omniscient.

Our phones are full of data-scraping apps, watching and listening to everything we do online. These beacons we tote everywhere transmit digital pings and signatures at all times. Corporations and national intelligence agencies know that data mining is key to gaining power, but only if they find new AI powers of computation. "Just as according to Christianity we humans cannot understand God and His plan," says Harari, "so Dataism declares that the human brain cannot fathom the new master algorithms."[59] Supercomputers will invent new ways of seeing and knowing that our puny brains cannot now conceive. The age of faith ends as Big Data proclaims to mankind: "For as the supercloud of data is superior to human knowing, so are my algorithms predictive of all your ways and my insights higher than all your thoughts."[60]

More optimistically, our medical diagnostics will be processed by AI, by an omniscient digital doctor who can simultaneously

---

59 Harari, *Homo Deus*, 398.
60 A futuristic paraphrase of Isa. 55:9.

recall your entire medical history, examine your full genome, view your scans and X-rays and vitals through a grid of medicine's entire recorded history, and look for problems to scale into a list of the curative, palliative, and preventative. AI can now accurately decipher the various protein 3D structures found in the human body, detect mutations, and target diseases with never-before-imagined drug remedies. Perhaps in due time tiny nanobots will circulate in our blood to provide real-time updates and 24/7 medical monitoring. The physician of the future won't be someone we visit once a year for a checkup but someone who watches over our real-time dashboard of data to call us when something goes awry. AI will know our body so well that we may find ourselves humanized to a detailed level we never imagined. AI's advantages are extensive.

But self-learning digital minds will disrupt economies and job markets. Just about everyone watching the rise of AI agrees here. Some forecasts say that AI, and its merger with autonomous robotics, will erase about half of our jobs and make a vast proportion of humanity inconsequential to the economy (according to Kai-Fu Lee). Others warn that the only humans to retain value in the AI age will be those with the foresight to plug a digital interface into their brains (according to Elon Musk). Others say that AI will initially disrupt a lot of jobs but then lead to more jobs, and better jobs, for humans (according to Kevin Kelly). Will AI decimate our chances of finding work? That's doubtful. More likely we will learn to work with AI like we've learned to work in front of computers today. The jobs of the future will call for AI fluency, not unlike the fluency already needed to work well with coworkers.[61]

---

61 Kevin Kelly, "The Future of Employment with AI," youtube.com (May 27, 2020).

But for a moment let's play out the harshest forecast. Imagine that upgrades to AI robots continue until humans are made irrelevant. Imagine that within thirty years, half of humanity has been told that the market value of their skills is zero. Telemarketers, customer service reps, accountants, cashiers, radiologists, pharmacists, translators, proofreaders, data entry workers, warehouse workers, long-haul truckers, farmers, fruit harvesters, couriers, Uber drivers, cooks, waiters, bank clerks, travel agents, stock-exchange traders, sports referees, and pretty much all fast-food workers—all gone, replaced by advanced computers and automated robots. An AI job takeover would mark the most disruptive economic and technological shift in human history. The downsides look stark: mass unemployment, middle-class evaporation, and a wealth disparity never seen before in a developed country. What happens when our marketable expertise has no value? What happens when my neighbor doesn't need any of my skills? What happens when jobs disappear and we are left, not with leisure, but with purposeless despair?

It's hard to imagine the widespread disruptions that would result. A man who has nothing to offer a wife and family in the way of labor is deeply restricted in his capacity to care for them. Men will find it more difficult to meaningfully marry and raise families. When children can be gestated in artificial wombs, and when sexbots and virtual sex workers cater to a man's sexual appetite, a woman's capacity for love will be profoundly curtailed too. As humanity separates between a "small elite of upgraded superhumans" and an un-upgraded majority, we will see the rise of the "useless class"—a mass of people "devoid of any economic, political, or even artistic value, who contribute nothing to the prosperity, power, and glory of society," and who live in isolation, feeding off chemical jolts or electrobuzzes of some artificial happiness for a life spent inside the telepathic hallucination of virtual reality.[62] In such a scenario, the sanctity of life, now spoken to protect the unborn from abortion and the elderly from euthanasia, will be called on to apply to a much wider demographic spectrum—those who are shunned by a market that doesn't want them.

The forecasts are dire for those who cannot adapt to the changes. "Yes, intelligent machines will increasingly be able to do our jobs and meet our material needs, disrupting industries and displacing workers in the process," AI prophet Kai-Fu Lee says about what he forecasts to be a major disruption headed our way. "But there remains one thing that only human beings are able to create and share with one another: love."[63] When the market devalues your skills and you have no skills to offer others, you still have your own presence to give.

AI will bring therapy bots, nanny bots, and pet bots. AI systems will offer the virtual appearance of presence, but they will remain impersonal. Instead, in the AI age, it looks like we may be in a position to love less via skill and love more proportionately in presence. According to Lee, loving others will be the key to finding personal purpose and thriving in an age when AI has taken your job away. There is some truth here. But we must go further.

In a sense, Christ prepared us to find our calling in the age of AI—and in any age. Here's how Jesus defines what it means to truly thrive in the world, in Luke 10:25–28.

And behold, a lawyer stood up to put him [Jesus] to the test, saying, "Teacher, what shall I do to inherit eternal life?" He

62 Harari, *Homo Deus*, 330, 355.
63 Kai-Fu Lee, *AI Superpowers: China, Silicon Valley, and the New World Order* (Boston: Houghton Mifflin Harcourt, 2018), 198.

[Jesus] said to him, "What is written in the Law? How do you read it?" And he [the lawyer] answered, "You shall love the Lord your God with all your heart and with all your soul and with all your strength and with all your mind, and your neighbor as yourself." And he said to him, "You have answered correctly; do this, and you will live."

This is the universal constant, the formula for vibrant life—to be fully and truly alive in this age and the age of AI. These are the two "love commands." In parallel accounts in the Gospels, Jesus himself makes the summary.[64] Here in Luke, the lawyer does. Unfortunately, the scheming lawyer won't get the point, because he's looking for self-justification, and that's not possible. Justification is found in the substitutionary atonement of Christ alone, in his supreme act of love toward us.[65] Nevertheless, the lawyer is not stupid. He boils down the entire moral will of God into two categories: love God with all that you are, and love others as yourself. Then Jesus commends the summary. The young man is right. This is what it means to be fully alive.

Note the primary love command: treasure God with everything you are. Here is the primary human vocation each of us is created to experience. More than trying to find a place in the market economy, each of us is created to express a daily, heart-soul-strength-mind, holistic embrace of God.[66] This is what it means to be alive. Love is a response to seeing God's glory and goodness, and in the light of his

---

64 Matt. 22:37–40; Mark 12:28–31.

65 2 Cor. 5:21; 1 John 4:10, 19.

66 Or in the words of Piper: "Jesus' demand to love God with all our heart and soul and mind and strength means that every impulse and every act of every faculty and every capacity should be an expression of treasuring God above all things." John Piper, *What Jesus Demands from the World* (Wheaton, IL: Crossway, 2006), 82.

beauty, desiring nothing on earth more than him, cherishing him above even the most beloved father or mother or son or daughter. Faith acts to joyfully give over all our earthly assets, everything the market values, in order to buy a field that holds the priceless treasure of Christ. Whether the AI age will give or take away, living faith considers everything in this life as loss compared to the supreme worth of knowing Christ.[67] We love God with all our heart and soul and strength and mind—holistic expressions of how we treasure him and find in him all that we ultimately desire. A life given to love God like this will renew the purpose of the "useless class" when nothing else will. This is the personal calling on each of our lives. And nothing, not even the most dire AI forecasts, will stop it.

Out of the primary love command flows the second love command: love your neighbor as yourself. Demonstrations of love will change in the age of AI. Love calls for creativity. But the lesson remains true for us today. Right now, we are tech wealthy. But we are commanded not to let these riches make us conceited and self-satisfied. Instead, finding our security in God alone—and never on the soon-to-disappoint promises of tech—we see in our innovations a generous giver who richly gives us every new power and comfort to enjoy. He calls us to enjoy and to use his gifts for good in serving the needs of others. Selflessly using our tech to meet others' needs is how we store up treasures in heaven.[68]

Jesus readies us for the AI age by reminding us that giving our lives away is what it means to be fully alive. The techno-control and pseudo comfort that the age of innovation promises will never make anyone more alive. True life is for those who are alive to God's beauty. This reality was true in the first century, it was true in the

67 Pss. 34:8; 73:25–26; Matt. 10:37; 13:44; Luke 10:27; 14:33; John 6:35; Phil. 3:8.
68 1 Tim. 6:17–19.

Industrial Revolution, and it will be true for whatever changes AI brings to our lives. As the technological age promises to make you the center of your own universe, our love commands forbid it. Our greater purpose shapes the priorities of our lives from outside of us, not from inside the technium. The Christian's life, meaning, and purpose will always be shaped by the greatness of God and by the presence we offer others.

6. We make space for unnecessary innovation.

We should look skeptically on any innovation that doesn't serve the flourishing of humanity. And yet something remains unsaid. Because I also don't think utility answers for all the holy impulses at work in human innovation and exploration. When I behold the eye-catching materials of this earth (gold, silver, and diamonds), or the expansive matter in the cosmos (more than 99.99% of which seems to exist without consequence to any of our lives), I am not left with the impression that creation was made simply for utility. The wild diversity of creatures in the animal kingdom makes this same point.[69] So when I see the three brothers of Genesis 4, I see an ancient culture striving to feed itself by spearheading innovations in genetics (thanks to Jabal) and protect itself by making metal tools (thanks to Tubal-cain). But it hardly seems like the appropriate moment for the invention of music. Music seems inconsequential compared to cattle and swords. But as unnecessary as it may have appeared on the surface, the Creator himself taught Jubal to make instruments of strings and wind.

I am astonished that I live on a globe so richly infused with minerals, natural laws, music, and latent powers that can animate our ambitions. I cannot help but be amazed by the creativity and generosity

69 Job 38:1–41:34.

and brilliance of the Creator. And within the very act of creation itself, wisdom is personified as both a playful child and a master craftsman.[70] The resulting creation is a playground for creatures.[71] So I think it would be a misreading of the creation to assume that only functional discoveries are of any value. I think that creation contains a wide range of possibilities to make and to discover new things that are not entirely pragmatic in their first cause or final purpose.

All human innovation is rooted in imaginative possibility, coming from a spontaneous Creator who shows off his own imaginativeness in the beauty and expanse of his creation, which he has stocked with commodities far beyond necessities for us to survive. Human innovation is never limited to fixing problems. It's the expression of playful spontaneity, of joyful discovery unfolding in a conversation with a Creator who has suffused creation with both necessary and unnecessary glories.

The sum total of our innovations will always exceed our necessity. Study the history of innovations, and a lot of serious fixes find their origin in discoveries meant for vain purposes (like Botox, as we will see in a moment). Humans will always be motivated by pure aspiration and simple curiosity, things like space exploration. Perhaps Mars will become our new home to save us from this failing planet. Maybe not. Maybe on Mars we will simply discover more of the creation pattern of our God—and that would not be wasted effort. Or maybe on some other planet we will discover a new metal more stunning than gold—something we will excavate, take back to earth, marvel at, and experiment on. Maybe it will make our computers faster. Maybe it will make our bodies healthier. Maybe not.

---

70 Prov. 8:22–31. See Leland Ryken et al., *Dictionary of Biblical Imagery* (Downers Grove, IL: InterVarsity Press, 2000), 128.
71 Job 40:20; Ps. 104:25–26.

When we look at human innovation only for its pragmatic uses, we miss something important about human life in the expansive sandbox of created wonders. Because creation is in fact a sandbox. At the end of the day, we are only playing with the possibilities given to us, down to the manipulation of sounds and frequencies that resonate with us as music. What if this creation was intentionally designed to be larger than all our necessities combined? What if, as we explore the nonnecessities, the Creator is also pleased to instruct us in what can only be wonderfully unnecessary? How would we think differently about technological innovation if we first pondered that music was God's idea?

God forbid that our innovations take too serious a turn, that we become so small-minded that we merely seek material wealth or happy investors or bull markets, and in the process lose the essential playfulness of our creativity.

7. We show patience with poisons.

The world is potent with poisons. And we make or discover new poisons all the time. Technology makes the drink and it makes the drunk. Technology makes the drugs and it makes the addict. Technology cultivates latent powers within creation that are medically helpful and recreationally harmful, things like marijuana and opioids. Technology leads to the mountain of ammonium nitrate needed every year to enrich the farmer's soil. And that same mountain explodes and levels a whole city.[72]

As we discover the boundaries of creation, we find mystery. Creation is loaded with dangerous poisons that baffle us. But here's the mystifying reality affirmed in church history. Sin, Satan, and

---

72 See the Port of Beirut explosion (August 4, 2020).

the fall cannot add any materials into creation. So how do we make sense of deadly viruses and bacteria that were apparently present in creation from the very beginning?

Augustine steps in to help us consider the naturally occurring poisons inside the created order. In *The City of God,* he first argues for the essential goodness of creation, even the dangerous parts that make us scratch our heads. Every part of creation is good and worthy of study and research. Consider poison, he says. "It is deadly when improperly used, but when properly applied it turns out to be a health-giving medicine."

When first discovered, bacteria was considered a poison and an enemy of human life. But that view has changed over time. The neurotoxin botulism, deadly in spoiled food, is now used in Botox injections to smooth out signs of aging and wrinkles. And when Botox patients reported fewer headaches, scientists discovered that the bacteria injections cured chronic migraines.

But Augustine's list of poisons extends to murderous viruses too. (*Virus* is the Latin word for *poison.*) Viruses also appear to serve the greater good, as they initiate important micro-evolutionary adaptations through a process called horizontal gene transfer. Viruses are gene hackers, often for good. In the words of one geneticist: "When bacteria were first discovered, few dreamed that they played such a critical positive role, as we now know they do in ecology—and that the same is also evidently true of viruses."[73]

Within creation we will find deadly *poisons* and—maybe we can go so far as to say—deadly *pathogens.* We may want to dismiss these discoveries as only destructive (I'm thinking specifically of anthrax and Ebola, but now also of the coronavirus). And there are

---

73 Jerry Bergman, "Did God Make Pathogenic Viruses?" answersingenesis.org (Apr. 1, 1999).

viral breakdowns that are incredibly dangerous. But the world is filled with bacteria and viruses that make life possible. In a fallen world, humans will be plagued by viruses gone rogue. This is part of what it means to live inside a cursed planet, a futility, a "bondage to corruption" that Paul sets forth in Romans 8:18–25. And yet we should not dismiss even the most dangerous viruses, because they may be found to be introducing important changes in reprogramming our DNA so that we can adapt to this ever-changing world. "Poisons" are mysteries, and we should postpone their judgment, because further study may reveal their importance.

Additionally, says Augustine, there are other strong potencies in creation that we've known about for a while but are still trying to handle wisely. I can think of caffeine, tobacco, alcohol, marijuana, psychedelics, hallucinogens, narcotics, opiates, snake venom, uranium, and fossil fuels. Each must be handled with care. Even good things become poisons with overuse, he says. Too much food, drink, or sunlight can hurt the body. That does not mean that food and drink and sunlight are evil; it means that we are still learning from divine providence "to be diligent in finding out their usefulness or, if our mind and will should fail us in the search, then to believe that there is some hidden use still to be discovered, as in so many other cases, only with great difficulty." From poisons to sunlight, we are always discovering, from the Creator, how to best use this gift of creation.[74]

In the end, God governs viral outbreaks for his own wise purposes.[75] And God instructs us to mitigate their spread and to stop

74 Augustine of Hippo, *The City of God, Books 8–16*, ed. Hermigild Dressler, trans. Gerald G. Walsh and Grace Monahan, vol. 14, The Fathers of the Church (Washington, DC: Catholic University of America Press, 1952), 220–21.
75 A few biblical examples of viral diseases that God sent directly or indirectly include Lev. 26:14–16; Deut. 28:59–61; 2 Sam. 12:15; 2 Chron. 21:11–20; Job 2:7. Additionally

their destruction with vaccines. God sends the thistles in the field, and God instructs us toward farming practices and pesticides to kill the thistles. The Bible helps us reconcile the sovereignty of God on both sides. He has use for thistles, viruses, and poisons. We may pin the blame for a virus on an accidental lab leak or a fluke zoonotic mutation, but every infection is a wake-up call allowed *by* God, and every scientific cure is a gracious gift *from* God. And yet, warns Augustine, we must wait patiently, because one day we may discover a positive use for every poison in creation. Be patient with the poisons.

8. *We get comfortable with tech minimalism.*

In a fallen world, man often feels helpless and dependent. Technology is his response, "a collective revolt against the limitations of the human condition," a revolt against the unruliness of nature and a revolt "against the reality of our dependence on forces external to ourselves."[76] We are porous to forces outside of us; therefore we reach for technology as a protective buffer. We misuse technology when we wield it in the hopes of self-sufficiency and autonomous protection from nature. We should never be more cautious than with technologies that seal us off from the natural world, that noise-cancel nature from our electrified lives.

Over the past sixty years, Christians have called out worldliness in categories of sex, drugs, and rock and roll. But in the tech age, worldliness sneaks in the back door, hooded in pragmatism and the desire for control. Lusting for sovereignty over

---

important texts to wrestle with here include Amos 3:1–6; Isa. 45:1–7; Lam. 3:37–39; Mic. 1:12.

76 Christopher Lasch, *The Culture of Narcissism: American Life in an Age of Diminishing Expectations* (New York: Norton, 1991), 243–45.

life, through tech, is a prevailing manifestation of worldliness in our age.[77] Tech "progress" is often driven forward by human lust for power.[78] Man seeks "absolutized scientific-technological control."[79] Said another way, "Technicism is the pretension of humans, as self-declared lords and masters using the scientific-technical method of control, to bend all of reality to their will in order to solve all problems, old and new, and to guarantee increasing material prosperities and progress."[80] When we wield technology in unbelief, we display a Babel-like expression of the human desire for sovereignty.

In stark contrast, to be human is to be a human *being*, a creature ordered toward God himself. We find our bearings in orientation to him and to his will. Creaturely autonomy is a fantasy. God's providence over the world, his church, and our lives is reality. In the words of John Webster, God's providence is "that work of divine love for temporal creatures whereby God ordains and executes their fulfilment in fellowship with himself." Out of his love, God orders our lives toward his glorious presence to enjoy forever.[81] But this precious promise is lost very quickly in the age of innovation. "We don't think that way today," warns Webster, "because we generally take a technological image of ourselves. We are essentially what we manipulate, what we make of ourselves through the things we make and the choices we make and the patterns we make around

---

77 On this connection see Craig M. Gay, *The Way of the (Modern) World: Or, Why It's Tempting to Live as If God Doesn't Exist* (Grand Rapids, MI: Eerdmans, 1998).

78 Egbert Schuurman, *Faith and Hope in Technology* (Carlisle, UK: Piquant, 2003), 21.

79 Egbert Schuurman, "A Confrontation with Technicism as the Spiritual Climate of the West," *Westminster Theological Journal* 58, no. 1 (1996): 74.

80 Schuurman, *Faith and Hope in Technology*, 69.

81 John Webster, *God without Measure: Working Papers in Christian Theology, vol. 1, God and the Works of God* (London, T&T Clark, 2015), 127. For the best full-length treatment of this theme see John Piper, *Providence* (Wheaton, IL: Crossway, 2021).

ourselves."[82] As Christians we want more for our lives and our children's lives than techno-manipulation. The Spirit must orient us toward God as our highest good, so that we not only believe it, but live from a conviction that he really is our supreme treasure, now and forever.

But another low-hanging fruit beckons us, a perpetual temptation to gain control over our bodies. No doubt the future of health will include more wearables as we attempt to quantify and data-fy everything from our heart rates, glucose levels, step counts, mood fluctuations, sleep patterns, and any manner of analytical readings for personal productivity. Anything we quantify into data we will try to optimize. Much of this will be good. And we will see new medical advances promising to end aging. But perhaps Western culture will become so infatuated with health that we will make ourselves ill. That's the suggestion of Packer. "Dazzled by the marvels of modern medicine, the Western world dreams of abolishing ill health entirely, here and now," he said. "We have grown health conscious in a way that is itself rather sick, and certainly has no precedent—not even in ancient Sparta. Why do we diet and jog and do all the other health-raising and health-sustaining things so passionately? Why are we so absorbed in pursuing bodily health? We are chasing a dream, the dream of never having to be ill. We are coming to regard a pain-free, disability-free existence as one of man's natural rights."[83] This "natural right" is the bruised fruit of technological control culture.

Packer warns that in our attempts to stop aging and optimize health through all sorts of quantified tracking and body hacks,

82 John Webster, "Discipleship and Calling (Part 1)," lecture, Scottish Evangelical Theology Conference (2005).
83 James I. Packer, "Poor Health May Be the Best Remedy," *Christianity Today*, May 21, 1982, 14.

we may miss out on God's bigger purpose and plan for our lives. "God uses chronic pain and weakness, along with other sorts of affliction, as his chisel for sculpting our souls," he wrote. "Felt weakness deepens dependence on Christ for strength each day. The weaker we feel, the harder we lean. And the harder we lean, the stronger we grow spiritually, even while our bodies waste away."[84]

Obviously, we can escape from God's providence like a fish can escape water for a life in outer space. But we resist God by thoughtlessly grabbing for more life control. This is idolatry. We'd rather have a god we can readily understand, easily appease, and instantly command.[85] Idolatry is all about control. And tech, like in the age of handheld idols, puts in our hands tools and gadgets that give us the appearance of control. It's a mirage. Any confidence we have about what we are going to do later today, tonight, or tomorrow is an idolatrous arrogance if we think that we ultimately decide. We don't. Our lives are a spritz-mist in the desert that evaporates before hitting the ground.[86] We are vapors. We don't control our lives because we don't control the living God. He's entirely other-than-us. We are creatures of clay. Our techno-control over the variables of this world is an idolatrous illusion. Instead, we affirm with the psalmist that God has governed my destiny until now, he is the source of all I need today, and he holds my future secure.[87]

None of these tensions are new to the Amish people. Contrary to this age of technological control that governs so much of urban

84 Packer, "Poor Health May Be the Best Remedy," 16.
85 Hab. 2:18–20.
86 James 4:13–17.
87 Ps. 16:5.

life, the Amish have retreated to the country with an intentional, self-limited "rightness of scale," a manageability that restrains the size of their farms and their proximity to community, "an economy dependent upon limits strictly understood and observed."[88] They limit their dependence on "machine-developed energy," and by it, says Wendell Berry, "have become the only true masters of technology."[89] Tech mastery requires self-limitation.

The Amish separate from the world into small communities, largely technologically isolated, adopting only a sparse number of tools that benefit (and do not harm) the local community. Kevin Kelly spent a lot of time with the Amish, studying their habits and convictions. He calls them "ingenious hackers and tinkerers, the ultimate makers and do-it-yourselfers."[90] The Amish are aware of iPhones and computers. But they are tech-adoption minimalists by clear convictions, particularly these four, in the words of Kelly:

1. They are selective. They know how to say no and are not afraid to refuse new things. They ignore more than they adopt.

2. They evaluate new things by experience instead of by theory. They let the early adopters get their jollies by pioneering new stuff under watchful eyes.

3. They have criteria by which to make choices. Technologies must enhance family and community and distance themselves from the outside world.

4. The choices are not individual but communal. The community shapes and enforces technological direction.[91]

---

88 Wendell Berry, *Essays 1993–2017* (New York: Library of America, 2019), 645–47.
89 Wendell Berry, *Essays 1969–1990* (New York: Library of America, 2019), 327.
90 Kevin Kelly, *What Technology Wants*, 217.
91 Kelly, *What Technology Wants*, 225–26; formatting original. See also, Jacques Ellul's "76 Reasonable Questions to Ask About Any Technology," thewords.com.

We can learn something from these four lessons, but something even more foundational is at work in this community. The Amish approach to life and technology includes intentional inaction—*gelassenheit*—a yieldedness, a serenity, a letting be, a relaxing from promises of techno-control over all of life, in order to submit to God's will over whatever is to come.[92] They seek to preserve one of the fundamental facts of humanness: we are creatures under the providence of God.

Likewise, Christians will help restrain the adoption of certain technologies based on dangers to creation, nature, and physical health (perceived from general revelation), and, most importantly, Christians will resist the adoption of technologies based on spiritual factors (learned from special revelation). Prudence will allow us to benefit from the best advances while limiting the misuse that often comes along with the false promises of tech control.

Here's the challenge. The dilemma of the tech age is how to live minimally without innovating minimally. "To maximize our own contentment, we seek the minimum amount of technology in our lives," writes Kelly, who learned this lesson while living inside an Amish community. "Yet," he says, "to maximize the contentment of others, we must maximize the amount of technology in the world. Indeed, we can only find our own minimal tools if others have created a sufficient maximum pool of options we can choose from. The dilemma remains in how we can personally minimize stuff close to us while trying to expand it globally."[93] This is what the Amish have figured out: how to remain aware of the proliferation of innovations happening around them while adopting tech minimally and based on the health of the community.

92 Adam Graber, "Amish Technology," thesecondeclectic.blogspot.com (May 2011).
93 Kelly, *What Technology Wants*, 238.

The Amish have pulled off coordinated tech minimalism. We won't. My minimalism will not look like your minimalism. This means we have warrant to innovate more broadly than any one of our personal adoption decisions. We are not called to stifle all new tech but to live with enough trust in God's providential control to celebrate the tech wealth offered to us while also demonstrating God-centered contentment required for a life of tech minimalism.

9. We expect God to guide, squelch, and hack human technology at will.

The appearance of big tech as an autonomous power is a mirage. Tech cannot buffer us from God. He wields tech as he wants to. He hacks it for his own purposes. And his reign over the horrific evils of technology is nowhere clearer than in the Roman cross. An upright wooden post with a transverse beam, the cross was a showcase. The criminal was nailed down by three iron spikes, the cross was planted in the ground, and the spectacle was lifted up for all to see. The cross was designed to kill criminals, insurrectionists, and disobedient slaves, and to do so slowly by exhaustion and asphyxiation. The slow death was public torture, a billboard of intimidation to say to the culture: "Behold the fate of any fool who defies Roman rule and threatens social stability."

But this awful tool of torture doubled as the hinge on which all of God's redemptive plan turned. God created metalworking to serve man's ambitious toolmaking, and man created metal spikes to kill man. God patterned trees to serve man's ambitious building, and man used the lumber to invent crosses to destroy man. In this most evil moment in human history, God's entire plan took a decisive step forward. Through the exploitation of technology, man killed the author of life. But God governed the entire episode. By a cosmic paradox that will never be eclipsed, in naked shame the tortured Christ exposed all the forces of evil in their defeat.[94]

A similar storyline unfolded in Babel. The great city meant to unify humanity got hacked. Humanity got dispersed across the globe. The very end mankind sought to prevent by their technology was the end that came upon them. The global economy today is proof that God can sovereignly hack any of our technological intentions.

At the cross or in Babel, even in the hands of the vilest intentions of man, technology is never outside the subversive providence of God. This remains the case today as we watch technological progress build from the patterns and possibilities of creation. And because tech cannot operate apart from God's governing hacks, Christians are free from the crushing anxiety of fearing rogue technologists. A geneticist in China cloning humans or an engineer in California designing a new superhuman species—each operates only within boundaries set by a sovereign God who governs all things at all times, and who subversively limits and hacks innovation at will for his redemptive purposes.

10. We commit to wielding innovations in faith.

Man's first temptation came in the offer of artificial intelligence, a new super-human capacity to make decisions by himself.[95] The upgrade to omniscience would free him to autonomously self-rule his ethics, and it would mute God's voice. AI was a promise

94 Col. 2:13–15.
95 Gen. 3:1–7.

273

of godlikeness. Falling for the temptation, Adam and Eve didn't unlock superintelligence, but their choice did open our eyes to right and wrong and loaded us with the burden of making ethical decisions (leading to long books with lots of application, like this one).

This same glossy, red-apple promise of superintelligence lives on. Compared to supercomputers, our natural powers of input and output (I/O)—ears, eyes, nose, mouths, thumbs, human I/O hardwired to a brain—represent a flow of information at the speed of a snail. We will communicate more swiftly and draw from all the accumulated knowledge of our brains if only we can be connected to a computer. As the tech trajectory moves in this direction, AI will promise godlike powers and make Godward trust seem utterly prehistoric and ridiculous. So be it.

But we know that the tech adoption question ("Should I?") should be followed by another question: "Does this new innovation seek to fulfill me in ways only Christ can?" We are image bearers of another. Our highest potential is never self-defined; we are porous plastic. Or, more biblically, we are clay—clay to be shaped by another, clay to be shaped into the likeness of the potter. We are ever becoming something, spinning on this globe as clay on the potter's wheel, never remaining the same people we were a year ago.

Proportionately, however, we are more often shaped by tech culture than by the Spirit of God. Technology is our self-making. Even back in Babel, new technological possibilities offered us new ways of becoming. And that is especially true today. In identity shaping, technology is "the most powerful force that has been unleashed on this planet, and in such a degree, that I think it's become who we are," says Kelly. "In fact, our humanity and everything that we think about ourselves, is something we've invented. We've invented ourselves."[96] That's hard to disagree with.

Babel's bricks led to a tower. Our gadgets make us live according to technologically possible behaviors today. New innovations shape us into who we are and how we express ourselves. Technological possibilities absorb into us and become new functions of our cyborgified being, new possibilities that are not a physical part of us and yet also an intuitive part of us. As our technologies fold into our lives, they define our self-projection. Our technologies promise to give us control over our uncertainties, our blemishes, and our self-expressions. Our technologies shape our comforts, our projected image, our vocational success, and even our spirituality. Our technologies express our inner hopes and aspirations. Technological powers become something of us. Fluency with our tools is great. But they go wrong when our technologies elicit in us a grab for control over our lives or when we use them to ignore God's calling over our lives. In these ways we fall into the spirit of Babel, the "bid for self-achieved security on the basis of technological progress."[97]

Contrary to our hallucinations of tech security, Ecclesiastes talks often of "striving after wind," or better translated, "shepherding wind."[98] Our attempts to control this world are like shepherding a wind gust, the very definition of vanity. When we realize that we cannot control the world, we finally have a foundation for our joy rather than a drainage ditch for our happiness. Why? Because while we cannot control everything, God does.[99] We will never shepherd

96 Kevin Kelly, "Technology's Epic Story," ted.com (November 2009).
97 Gordon J. Wenham, "Genesis," in *New Bible Commentary: 21st Century Edition*, 69.
98 Eccles. 1:14, 17; 2:11, 17, 26; 4:4, 6, 16; 6:9.
99 Peter J. Leithart, *Solomon among the Postmoderns* (Grand Rapids, MI: Brazos Press, 2008), 168.

the wind. But we serve the God who controls the wind and rides it like a tamed horse.[100] We find our security in God alone.

11. We open ourselves to God-centered wonder over the gifts of technology.

Technological innovations have always attracted human awe because they disclose the cutting edge of human imagination and physical possibility. The original seven wonders of the world were all feats of engineering: pyramids, statues, towers, and temples. Our innovations are even more captivating because we can see the rapid change, even in the course of one lifetime. If you were born on December 17, 1903, the date of the Wright brothers' first successful flight, you could have witnessed the first fighter jet flight at age thirty-seven, bought a ticket for a commercial jetliner at the age of forty-eight, seen the first rocket launch into space at age fifty-three, watched the moon landing on TV at age sixty-five, and attended the first space shuttle flight at age seventy-seven.

Tech changes are especially obvious in this country. America was first to fly, first to walk on the moon, first to split the atom, and first to drop the bomb. Tech marvels have always awed the world.[101] And the global race to self-driving cars, domestic robots, 3D-printed manufacturing, cashless cryptocurrency, and quantum computing is well underway. Who will wow us next?

"I am willing to bet that in the not-too-distant future the magnificence of certain patches of the technium will rival the splendor of the natural world," Kelly writes. "We will rhapsodize about this or that technology's charms and marvel at its subtlety. We

---

100 Pss. 18:10; 104:3.
101 See David E. Nye, *American Technological Sublime* (Cambridge, MA: MIT Press, 1996).

will travel to it with our children in tow to sit in silence beneath its towers."[102] It's a safe bet. But only the shortsighted parent will rhapsodize over the glory of tech or the glory of robots or the glory of man.

Back in the fall of 1888, Charles Spurgeon heard recorded music for the first time. It's hard for us to imagine this; recorded music has *never not been* part of our lives. But the innovation was brand-new for Spurgeon. In 1888 Jubal's tech tree leapt forward in progress. "I sat yesterday with two tubes in my ears to listen to sounds that came from revolving cylinders of wax," Spurgeon said. "I heard music, though I knew that no instrument was near. It was music which had been caught up months before, and now was ringing out as clearly and distinctly in my ears as it could have done had I been present at its first sound. I sat and listened," said Spurgeon, "and I felt lost in the mystery."

But then Spurgeon wondered out loud. Why are we not lost in the mystery of the gospel? Why do shiny new innovations more easily capture our wonder? Why was it that as electric lights began illuminating London churches, the glory of Christ began dimming into the intellectual skepticism of the age? The glory of the gospel is more wonderful than electricity and the radiance of engineered lighting. Spurgeon said, "In the gospel of the Lord Jesus, God speaks into the ear of his child more music than all the harps of heaven can yield. I pray you, do not despise it. Be not such dull, driven cattle that, when God has set before you what angels desire to look into, you close your eyes to such glories, and pay attention to the miserable trifles of time and sense."[103]

102 Kelly, *What Technology Wants*, 325.
103 C. H. Spurgeon, *The Metropolitan Tabernacle Pulpit Sermons*, vol. 34 (London: Passmore & Alabaster, 1888), 531–32.

Angels don't bend down in awe of Silicon Valley. The angels kneel in awe to study the glories and agonies of Jesus Christ.[104] So should we. Only a dull beast would idealize technological progress and the Gospel of Technology. Rather, when we are oriented properly toward the soul-satisfying Creator, technology is no longer idealized into a god. Only then can our technologies become as mysterious and marvelous as the first recorded music.

Few better celebrated God's kindness in human innovation than G. K. Chesterton. He once wrote:

> It was the glory of the great Pagans, in the great days of Paganism, that natural things had a sort of projected halo of the supernatural. And he who poured wine upon the altar, or scattered dust upon the grave, never doubted that he dealt in some way with something divine; however vague or fanciful or even skeptical he might be about the names and natures of the divinities. Wine was more than wine; it was a god. Corn was more than corn; it was a goddess. . . . They were not satisfied with realism, because they never quite lost the sense of something more real than realism. They were not content to call a spade a spade, because it was almost always a sacred spade; not only when it dug the graves of the dead, but even when it dug the garden to grow fruit for the living.[105]

In the ancient world, the material world was entirely porous to the divine. Everyday things held spiritual meaning. In the tech

---

104 See 1 Pet. 1:10–12. Point inspired by Stephen Charnock, *The Complete Works of Stephen Charnock* (Edinburgh: James Nichol, 1864–1866), 4:70.

105 G. K. Chesterton, *The Collected Works of G. K. Chesterton, vol. 36, The Illustrated London News, 1932–1934* (San Francisco: Ignatius Press, 2011), 81.

age, the material world has a rock-hard graphene resistance to the divine. The material world is divorced from the spiritual world. Atheism dominates. Sinners do their best to ignore God and live as if he is fiction. And yet the Creator is here, giving us new industries, creating new industry leaders (*bara*), and abundantly blessing us with new tools. He gives in order for us to see his glory manifested. God's will is not complex. Everything he does is for one end—himself.[106] And that includes us and all of our innovations. So why did God load all the potentialities in creation? Why did he inspire us and teach us arts, agriculture, metallurgy, genetics, and music recording? He is showing off his glory so that our hearts will worship him.

Tech-age man will go on being awed by quantum computing and whatever scaled acceleration we make next. "Speed is the form of ecstasy the technical revolution has bestowed on man."[107] But this rush-induced ecstasy pales to joy in the giver of the technological revolution. True technological awe is not focused on accelerated speeds or new shiny gadgets. These are merely gifts from the giver. By our God-centered worship for these wonderful tools, we break the stranglehold of the tech age and its pursuit of a godless, artificial, and autonomous security. We know better. We see divine generosity in the thousands of innovations we use daily.

12. We refuse to concede our Sabbath rest to the demands of technology.

We may always work with machines, but we must never become machines. The human body is remarkable in similarities to an

106 Petrus Van Mastricht, *Theoretical-Practical Theology, vol. 2, Faith in the Triune God,* trans. Todd M. Rester (Grand Rapids, MI: Reformation Heritage, 2019), 297–300.
107 Milan Kundera, *Slowness: A Novel,* trans. Linda Asher (New York: HarperCollins, 1997), 2.

efficient machine, but we are physical beings with finite limitations and eternal souls. We process slower than supercomputers, but that's no flaw in our design.

Ever since the miner created the first vocation set free from the circadian rhythms of day and night, man has been tempted to overwork. We are always tempted to be something more than human. AI pushes this desire to new heights, calling for humans to begin intellectually keeping pace with the warp speed of machine learning. Today Elon Musk wants us to think, "We are literally a brain in a vat. The vat is your skull. Everything you think is real is an electrical signal."[108] And if you don't agree with his human-computer model, Musk gives an ominous forecast. "Under any rate of advancement in AI, we will be left behind by a lot," he said of humans. "The benign situation with ultra-intelligent AI is that we would be so far below in intelligence, we'd be like a pet, or a house cat. I don't love the idea of being a house cat."[109]

In order to compete with superintelligence, we must become more than brains in a vat. We must become cyborgs, brains augmented with high-power computing capabilities. To remain relevant, humans must adapt to the warp-speed advances of our technologies. Humans must identify computermorphically. We must become machines.

If we fail to resist this techno-tyranny, we *will* become cyborgified machines. We will live like Charlie Chaplin in his 1936 silent film *Modern Times*, in the scene when his frantic bolt-tightening pace is too slow and he gets sucked by a conveyor belt down into the gears of the machine, twisted and curved like a human chain. We are

108 Elon Musk, @elonmusk, Twitter (Dec. 12, 2019).
109 James Titcomb, "Elon Musk: Become Cyborgs or Risk Humans Being Turned into Robots' Pets," telegraph.co.uk (June 2, 2016).

not machines. Our relevance is not determined by our unstopping output. But man has always been tempted to work like a machine, even back in the age of the steam engine.

From the pulpit Spurgeon once admitted: "I am always ready to try a new machine."[110] He was an early tech adopter and loved new gadgets. I can imagine no better Londoner to hear Edison's first recording. His own preaching style was so radical that a newspaper editorial cartoon once satirized him preaching while sitting atop a rushing express train.[111] Spurgeon was innovative, fast, and revolutionary, but he knew how to pull the brakes. He used the steam engine as a metaphor of caution to warn his church of this tendency for humans to morph into the image of their machines. "Ours is not a religion of mechanics and hydrostatics: it is spiritual, and must be sustained by spiritual means."[112] One hundred days before the golden spike was driven to connect America's First Transcontinental Railroad, opening new doors for express travel on the rails, Spurgeon preached this concern: "In these days, when everybody travels by express and works like a steam-engine, the mental wear and tear are terrible, and the advice of the Great Master to the disciples to go into the desert and rest awhile is full of wisdom, and ought to have our earnest attention."[113] Techno-dehumanization is older than sliced bread, for tech has always tried to tempt us with the stale bread of anxious toil.[114]

---

110 C. H. Spurgeon, *The Metropolitan Tabernacle Pulpit Sermons*, vol. 26 (London: Passmore & Alabaster, 1880), 392.
111 See the satirized sketch of Spurgeon titled "The Fast Train" at the British Museum.
112 C. H. Spurgeon, *The Metropolitan Tabernacle Pulpit Sermons*, vol. 13 (London: Passmore & Alabaster, 1867), 231.
113 C. H. Spurgeon, *The Metropolitan Tabernacle Pulpit Sermons*, vol. 15 (London: Passmore & Alabaster, 1869), 62.
114 Ps. 127:2.

In the digital age, man is told to become a hyperprocessor like a computer. In the Industrial Age, man was told to become hyperkinetic like a factory. And in the age of the steam engine, man was told to maintain the hypertorque of unstopping pistons. The message of fear was the same: "Accelerate or be run over." In the age of steam, machines, and computers, the church reminds the world of the Sabbath rest.

For all its good, the technium will never understand the Sabbath, nor will it understand anthropology 101, why humans are not angels or animals or robots or machines or computer processors. Preserving the nature and purpose of man will be the work of the church for a long time to come. We slow down. We stop. We let the boiler tank of marketable activity stop and cool down. Our day of rest reminds us and the world that we are humans made for something greater than hyperaccelerated, nonstop computation and production.

13. We use innovation best when it serves our communion with God and others.

Technological innovation often feeds self-centered power, and that self-importance "always wins out over the love which seeks the well-being of the other," writes Egbert Schuurman. "In a technicized culture, communal ties are readily cut and replaced by technical or organizational relations. Love dies; empathy and sympathy and contact with the other disappear. Estrangement and loneliness increase."[115] The Amish saw this problem decades before same-day delivery from Amazon Prime rendered neighbors unnecessary.

Technology distances us from others and can distance us from God too. But it shouldn't be this way. In the early 1970s Victor

---

115 Egbert Schuurman, *Faith and Hope in Technology* (Carlisle, UK: Piquant, 2003), 101.

Ferkiss wrote, "If we can create a society of technological men who are the conscious masters rather than the unthinking slaves of their technologies, then perhaps technology will be able to fulfill its mission, to provide a base of physical security from which we can explore ever more intensively what it means to be human—what men can and should be." The important new frontier of exploration in the future is not to the utter ends of the universe but into our souls. "We are already well on our way to conquering the outer space of physical nature; our real task, though, still lies ahead—the conquest of inner space and the development of our fullest spiritual potential."[116] Colonizing Mars is not man's greatest challenge. In a techno-materialistic world, new innovations are inevitable. The real challenge is in seeking spiritual flourishing.

In the tech age we are still called to communion with God. Our gadgets and powers are enchanting, but they are not greater than meeting with the God of the universe. Spurgeon considered electricity to be a spiritual force more than a material force, since it had, like the spirit world, broken free from "the chains of time."[117] Nevertheless, even if electricity travels at more than 200,000 miles per second, "prayer travels faster," he said.[118] Even in the age of fiber-optic wires whizzing data at speeds measured in terabits per second, communion with God via the gift of prayer is an instant, superconnected power given to each of us.

Christian prophets of the past imagined a beautiful world where the necessities of labor were minimized to give us more time in

116 Victor C. Ferkiss, "Technology and the Future of Man," *Review and Expositor* 69, no. 1 (1972): 54.
117 C. H. Spurgeon, *The Metropolitan Tabernacle Pulpit Sermons*, vol. 17 (London: Passmore & Alabaster, 1871), 499–500.
118 C. H. Spurgeon, *The Metropolitan Tabernacle Pulpit Sermons*, vol. 61 (London: Passmore & Alabaster, 1915), 525.

communion with God. Jonathan Edwards could not imagine AI, but he did appreciate human ingenuity and expected future tech to expand time for leisure and contemplation. The expanse of our video platforms proves how much leisure time we now enjoy. But long before Netflix, Edwards predicted this growing margin and set forth a vision as he meditated on man's tech future. Contemplating a gadget in his office in 1725, Edwards foresaw an age of innovation that would empower communion with God and connect the global church in real time. The optimistic postmillennialist penned this prediction:

> 'Tis probable that this world shall be more like heaven in the millennium in this respect, that contemplative and spiritual employments, and those things that more directly concern the mind and religion, will be more the saints' ordinary business than now. There will be so many contrivances and inventions to facilitate and expedite their necessary secular business, that they shall have more time for more noble exercises, and that they will have better contrivances for assisting one another through the whole earth, by a more expedite and easy and safe communication between distant regions than now. The invention of the mariner's compass is one thing by God discovered to the world for that end; and how exceedingly has that one thing enlarged and facilitated communication! And who can tell but that God will yet make it more perfect; so that there need not be such a tedious voyage in order to hear from the other hemisphere, and so the countries about the poles need no longer to lie hid to us, but the whole earth may be as one community, one body in Christ.[119]

119 Jonathan Edwards, The "Miscellanies": (Entry Nos. A–z, Aa–zz, 1–500), ed. Thomas A. Schafer and Harry S. Stout, corrected ed., vol. 13, *The Works of Jonathan Edwards* (New Haven, CT: Yale University Press, 2002), 369.

I love that phrase: "by God discovered to the world"—a whole theology of technology is loaded into those six words.

Face-to-face conversations and body-to-body hospitality will remain virtues of the Christian life in the technological age. But technologies can also bridge spatial gaps and join us together in profound ways. Edwards meditated on the mariner's compass and from there predicted an age of innovation that would span the globe and overcome physical distances. He would be stunned to see his predictions come to fruition in the Internet, Twitter, Facebook, Instagram, YouTube, and live-streaming video, a myriad of ways of joining the global church digitally in real time.

Compared to history, we should celebrate technology because it saves us time at work, gives us more time with God, and unites us in global fellowship with God's people. These standards of gratitude hold fast in the tech age.

14. We submit our innovations to the wisdom that subverts the powers of man.

To my surprise, Christians are often threatened by the powers of technology, as if the powers of the technium should intimidate the weak. The tech age says that power is perfected in greater power. Power is perfected in autonomous robots. Power is perfected in proprietary algorithms. Power is perfected in AI. Power is perfected in genetic enhancements. Power is perfected in techno-humanism, transhumanism, and posthumanism. But God has always flipped the script on the world's power-lust because God's strength is most perfectly displayed in weakness.[120] The big tech overlords with all their money and coercive power

---

120 2 Cor. 12:9.

are thwarted by the world-altering power of daily Christian obedience.

In the cross of Christ, God thwarted the superhuman ingenuity of man, and he did it by foolishness. Jesus Christ is tech-culture foolishness personified. In Christ the world sees only a crucified fool. But that fool is the wisdom and power of God to make the arrogant wisdom and power of techno-man look like the folly it really is.[121] Satan still tyrannizes the earth and veils the hearts of sinners with man-made powers in order to blind them to the glory of Christ.[122] But Satan's power over this world has also been decisively cut in the weakest and most foolish way imaginable, in the cross.[123] The crucified and resurrected Christ is "the power of God and the wisdom of God" (1 Cor. 1:24). He is the center of our existence. He "became to us wisdom from God" (1 Cor. 1:30). All the hidden treasures inside this earth cannot compare to Christ, in whom "are hidden all the treasures of wisdom and knowledge" (Col. 2:3).

If you want to plumb the depths in order to discover the purpose or meaning of life, you won't find it in a deep mine shaft. You won't find it in endless innovation. We find divine wisdom by plunging into the endless depths of the person of Christ.

The power of tech will challenge our base of authority. In the greatest challenges of our age, where will we turn? In life's greatest challenges, some will trust in chariots, some in horses, and some in search engines. "When you talk about God and religion, in the end it's all a question of authority," Harari says, "What is the highest source of authority that you turn to when you have a problem in

---

121 1 Cor. 1:18–2:16.
122 2 Cor. 4:4.
123 John 12:31–32.

your life? A thousand years ago you'd turn to the church. Today we expect algorithms to provide us with the answer—who to date, where to live, or how to deal with an economic problem."[124] Will we trust autonomous intelligence to save us? Will supercomputers become our ultimate authority? Will Google become our god?

Tech powers unleashed in the nineteenth century captured Spurgeon's attention. He knew that greater wonders were to come, along with new challenges to Scripture's authority. He said:

> Was there ever such a century as the nineteenth? Was there ever such a period of time since the world began? What is there that we are not doing? Lighting ourselves by electricity, speaking by means of the lightning, traveling by steam—what a wonderful people we are! Yes, yes; and we are going to do much greater things than these, no doubt; and many matters, which are now reckoned as mere dreams, will probably become accomplished facts in a few generations. But after these marvels have all come and gone, the words of our Lord Jesus Christ will still abide, they will not pass away. Fashion follows fashion; systems succeed systems; everything beneath the moon is like the moon, it waxes and wanes, and is ever on the change. But come whatever change there may, even if the human race should reach that wonderful development which some prophesy for it, yet still, the words of our Lord Jesus Christ shall not pass away. And when the greatest alteration of all shall take place, and this present dispensation shall come to an end, and all material things shall be consumed with fire, and be destroyed, yet, even then, there shall remain, above the ashes of the world, and all that is therein,

124 Olivia Solon, "Sorry, Y'All—Humanity's Nearing an Upgrade to Irrelevance," wired.com (Feb. 21, 2017).

the imperishable revelation of the Lord Jesus Christ, for, as Peter says, "'the word of the Lord remains forever.' And this word is the good news that was preached to you" (1 Peter 1:25).[125]

The technium's only power is in the kinetic ebb and flow of new innovation. The system imagines, reaches, and brings forth new possibilities into material reality by following along the patterns made available by the Creator. The technium will appear to grow stronger and even unstoppable. But the paradoxical reality—which Christians know, and Silicon Valley never will—is that man is never weaker than when he looks the strongest. God limits the relative power of technologists through subversive weakness, as the Son and Spirit operate in the lives of ordinary Christians and churches. God's power works through Christians inside tech centers and outside tech centers, through Christians inside tech corporations and outside tech corporations, through ordinary-looking saints who seek to serve their God by fearing him, obeying him, and trusting the eternal word that cannot perish.

Tech Is Never Enough

Technology may appear strong, but it is weak—too weak to satisfy the desires inside of humanity. When Carl Sagan finished mapping out a multiplanetary possibility for the survival of humanity, he warned that we might be safe from the demise of this planet, but we would never be safe from ourselves. We humans carry within us a propensity for self-destruction—on earth, Mars, or whatever planet we choose next. "If we become even slightly more violent, shortsighted, ignorant, and selfish than we are now," he warned,

---

125 C. H. Spurgeon, *The Metropolitan Tabernacle Pulpit Sermons*, vol. 45 (London: Passmore & Alabaster, 1899), 398; slightly edited for readability and KJV replaced with ESV.

"almost certainly we will have no future."[126] We are selfish. We are not easily satisfied, certainly not by our technologies.

Ephemeral things are shinier than eternal things, and new innovations are always more alluring than old insights. Microprocessors and smartphones expand what it means to be human. Tools teach us more about ourselves and help us express ourselves more fully. Tech tools are not like wrenches and hammers we use for limited purposes; they are tools of self-discovery and self-expression. Our most powerful tools expand our lives, our aspirations, and even our loves. But as we embrace new possibilities, we are also left with a huge new dilemma.

The spiritual dilemma of the tech age is deep because our modern economy is built on the false promise that new innovations are the key to satisfying the heart's longings. "If we are honest, we must admit that one aspect of the ceaseless upgrades and eternal becoming of the technium is to make holes in our heart," Kevin Kelly admits. We are made discontent by design. "One day not too long ago we (all of us) decided that we could not live another day unless we had a smartphone; a dozen years earlier this need would have dumbfounded us," he writes. "Now we get angry if the network is slow, but before, when we were innocent, we had no thoughts of the network at all. We keep inventing new things that make new longings, new holes that must be filled." Technological discontent is not dehumanizing, he says; it's human-enlarging. Our new tools make us more of who we are. They expand us. But by expanding us, they pull open more new holes to be filled. "The momentum of technologies pushes us to chase the newest, which are always disappearing beneath the advent of the next newer thing,

126 Sagan, *Pale Blue Dot*, 329.

so satisfaction continues to recede from our grasp." So what is the solution? In the end Kelly turns to the technium to "celebrate the never-ending discontentment that technology brings," because this "discontent is the trigger for our ingenuity and growth."[127]

New tech innovation is triggered by the discontent within us. We are never satisfied, always searching, always willing to adopt new tech in the pursuit of self-fulfillment that never comes. It's heart-breaking to see an honest man reckon with the disappointments of tech. Every new gift of innovation promises to make more of us, but in that promise we get poked with new holes of neediness that must be perpetually filled with more tech putty. More tech adoption means more needs, more holes, less fulfillment, and more need for more tech adoption. Tech stocks feed off this dissatisfaction, but for the human soul this is a nightmarish projection from the pages of Ecclesiastes. Innovations leave us dry because they intentionally ignore God. And any scientific or technological endeavor that attempts to leave God out "becomes its own opposite, and disillusions everyone who builds his expectations on it."[128] The disappointments of the technoverse prove this over and over. We've now seen this tech disillusionment twice in history, at the close of the nineteenth century and inside twenty-first-century Silicon Valley. This disenchantment will always be there. Two concurrent evils always subvert man's happiness: first, forsaking God as our all-satisfying "fountain of living waters," and, second, replacing him with some newfangled human-engineered promises that cannot rise higher than "broken cisterns," tanks full of unpatchable holes that can hold no water (Jer. 2:13).

---

127 Kevin Kelly, *The Inevitable: Understanding the 12 Technological Forces That Will Shape Our Future* (New York: Penguin, 2017), 11–12.

128 Bavinck, *Wonderful Works of God*, 4.

When we seek happiness in the latest tech feat of man, we must first assume (knowingly or not) that God is not enough. Our latest gadget promises to complete us. But God knows that it won't. We are eternal souls who cannot be satisfied in the ambitions of man. God knows this, and he subverted the false promises of the Gospel of Technology from the start. You can fill your heart with man-made replacements for God, but they will never be enough. You can chase after the next tool or the newest innovative power or the newest augmentation or for the next frontier in space exploration. But if you are doing these things to satisfy your heart, you'll be stopping holes with Silly Putty.

Technologies are wonderful. The potent computer chip changes everything. The power of digital cameras is spellbinding. The smartphone is stunning. The Internet that joins together Christians from across the globe is remarkable. Space travel that expands what we know about the universe is breathtaking. Medical advances, like the end of polio and the end of cancer and the end of dementia and the end of genetic defects—should we see these victories—would be astounding, and we would give God all the glory for creating minds to address these problems. But Christians must always return to the issue of trust. The same bucket of tar can be used to build our trust in God or to build towers of unbelief.

Wise living in the tech age is not settled by Christians who ignore material possibilities, nor by the technologists who dismiss spiritual wonders. Wisdom begins in fear and is expressed in gratitude. Can I—in good conscience—thank God for an innovation? The ethics of what is permissible or forbidden is rooted in gratitude. "For everything created by God is good, and nothing is to be rejected if it is received with thanksgiving, for it is made holy by the word of God and prayer" (1 Tim. 4:4–5). This holds true for barbecue and marriage and smartphones and medical advances. If we can honestly thank God for it, we can adopt it. God-centered gratitude gives us faith to see that only Christ can fill the holes in our souls. Christ is the secret to thriving in the age of AI, autonomous robots, and anti-aging advances. Joy in Christ teaches us to be thankful for the innovations we need and content without the ones we don't.

Nine Tech Limiters

The pinnacle of our technological age sounds a lot like Babel and Babylon—reaching a place where nothing can limit or restrain us. "One day our knowledge will be so vast and our technology so advanced," predicts Harari, "that we shall distill the elixir of eternal youth, the elixir of true happiness, and any other drug we might possibly desire—and no god will stop us."[129] But believers know, by faith, that the Creator *can* stop us. He *does* stop us. God governs his creation with governors, or valves, to intentionally limit human innovation and its possibilities.

God is inside Silicon Valley. He is at work inside every tech center on the globe. He is at work in at least nine areas I can see. I'll call them "nine tech limiters."

1. Creational limiter. The Creator regulated what is possible in this material creation. By his intentional patterns he limited the world's natural resources and their abundance or scarcity. He restricted those material possibilities even further by natural laws.

2. Vocational limiter. The Creator slows or accelerates human innovation like a thrust lever in a cockpit. To his own intended

129 Harari, *Homo Deus*, 202.

volume, God creates (*bara*) new innovators and wielders of innovation within any given era of history—and this includes the ravagers, the ambitious, and the virtuous.

3. Cross-cultural limiter. The Creator initiated cultural tensions within tech adoption to help protect humanity. By multiplying all the cultures of the world at Babel, God set in place a global friction to limit what innovations are adopted in any given society.

4. General-revelation limiter. The Creator's voice instructs through the created order. As in a catechism for all to read, we hear how our actions impact the creation for its good or harm.

5. Interruption limiter. God wields direct power to intrude on human tech. By his sovereign inbreaking he can thwart any technologist or hack human innovation for a greater purpose.

6. Spiritual limiter. God reveals his specific intentions for creation to the church. Through the gift of special revelation, through his word, God gives wisdom to his people to spiritually flourish and discern their way through the complexities of the tech age.

7. Weakness limiter. God overturns the power structures of man. By the foolishness of the gospel and the simple obedience of ordinary Christians, God intentionally subverts the seeming unthwartable power of big tech in order to accomplish his ends.

8. Heart limiter. God exposes the insatiable appetites of the human soul. He designed all human technological advances to expose the soul's vast spiritual need, without any possibility of that growing thirst to be satisfied by anyone or anything but himself.

9. Hard-stop limiter. God puts an end to human technology. In the ashes of Babylon, God reveals his final commitment to put an end to human techno-idolatry and urban rebellion so he can make way for something greater.

By these nine limiters, God ultimately controls the start and pace and end of all human innovation. Whether we trust the Creator in these areas will determine how we face an uncertain technological future: with fearful reserve or with hopeful freedom. Trust makes the difference.

Our technological age will continue to showcase human aspiration and achievement. The theme of our age is man, man—all about man. To thrive in such an age requires more faith in God, not less. In the face of the never-before-witnessed power of man, our God must become even more real, even more there, even more aware, even more in command of all things. And that is the God we serve. By faith we know that God governs every active agent. No free-will AI conscience falls outside of God's providence because God creates AI's creators. The ancient ravager was an active agent, indeed an autonomous acting agent, an AI ravager. In our future, we may see autonomous military bots, killing machines that will decide who dies based on internal computation and algorithms. But autonomous robots, like autonomous humans, are never actually autonomous. Whether made of flesh or steel, no acting agent can escape God's governance.

All of mankind once coalesced and concentrated their engineering power in a tower to thwart God. It failed. God spoiled their aims without a sweat, unleashing all the cultures of mankind. When mankind conspired again with all angels and demons and political powers to kill God, they pinned the Son of God to a cross. But it was a baited trap. In the cross Christ defeated the devil, death, and sin. And when Babylon unites humanity into a global superpower, God will end it. Man is never in control. Frankly, if God did not have the power to thwart human innovation and engineering, what hope would we have for the future? We could only expect an unabated future of techno-tyranny. But human innovation never sits in God's throne.

In the church, fear is winning out over faith when it comes to technology. God feels distant from tech culture when the god in our heads seems outmatched by the power of man. We must stop living in this theological fiction. We must return to the God of Scripture so we can trust his providence over the material universe and over every turn in the history of human scientific and techno-logical change. This turn will protect us from the despair of seeking to accomplish what the agnostic technologist already attempts in wielding human control over creation. If the god in our heads can-not stop man's tech, maybe we should? This is a disastrous lack of faith. As the technium grows beyond the church's control, we must trust in the living God of the universe who governs all things. Only with these governors in place can we look optimistically over the unfolding drama of technology before our eyes.

By faith, I eagerly await what's next on the horizon as the glorious patterns of creation offer us new possibilities we never imagined. My life is filled with centuries of tech advances, and I am eager to see the new tech of the future that exists beyond my present imagination. The unfolding story of technology is a roller coaster we cannot stop, a train we are all strapped into as the car inches its way up the first peak. While things are relatively calm now, we expect to soon plunge into a fury of new ideas and possibilities that will be unfamiliar. It's too late to jump out. In such a scenario, I'm advocating for the faith to ride it out, to ride without fear, to ride like people who trust in the God who reigns over all things and loves us to the point of shedding the blood of his own Son for us to prove that he will give us everything we need in the tech age.[130]

It's going to be a wild ride. It won't always be comfortable. We will overreach. We will attempt too much. We will make mistakes and maim ourselves along the way. We will always be in need of correction. But by faith we can rest assured that the technium will never escape God's nine limiters. So I'm optimistic—not optimistic in man, but optimistic in the God who governs every nut, bolt, chain, and seat belt in this wild technological carnival ride.

The Crescendo

We face the tech age with sobering questions. What if our tech age dehumanizes us by vanity? What if we become so good at amassing comforts that we become terminally bored? What if we become so stimulated by sexbots we become impotent? What if we modify our bodies until we no longer long for glorification? What if AI systems augment us with all the known knowledge of mankind until we never again experience mystery or wonder? What if we become so medically buzzed on electrical stimuli or artificial chemicals that we can no longer feel genuine human emotion? What if we grow so powerful that we live inside an impervious isolation that pushes away everyone else, and God too?

130 Rom. 8:32.

When Einstein called science "the most precious thing we have," he sold us short.[131] Christ is the most precious thing we have. And it's in the light of Christ that we see science more clearly. In him we see its purpose, its goal, its designer. For in all things, including science and technology, Christ is the Alpha and Omega, the first and the last, the beginning and the end.[132] Christ existed before all things because he made all things. Now he sustains all things. By him "all things were created, in heaven and on earth, visible and invisible," and "in him all things hold together." Christ is the creator, preserver, and end of creation, for all things "were created through him and for him" (Col. 1:15–17).

The glory of Christ is the epicenter of Christian flourishing in the tech age and in any age. His glory is the trans-technological priority for the gaze of our lives, the solid ground for our minds and wills and souls and hearts in whatever social changes have come or will arrive in the future. The church will continue to exist on this earth as a refuge for those who inadvertently break themselves under all the false, dehumanizing promises of self-redemption and self-security in the biological manipulation of the body and in the cyborg-like augmentation of our physical and cognitive powers. The false promises of superpowers and superknowledge will be too much for many to resist. But those promises will disenchant over time, and sinners who break themselves will seek deliverance. There will always be hope for broken souls and broken bodies, even self-broken ones, in Christ. Tech will extend life, but it cannot stop death. In Christ we are delivered from death into a community that is being transformed from one degree of glory to another in ways no lab will copy.[133]

131 Cited in Banesh Hoffmann, *Albert Einstein: Creator and Rebel* (New York: Penguin, 1972), *v.*
132 Rev. 22:6–21.
133 2 Cor. 3:18.

Christ's supremacy over all things means that Christian flourishing does not hinge on my adoption or rejection of certain technologies. It hinges on my heart's focus on the Savior. This will be true across the spectrum of tech-adopting Christians and tech-rejecting Christians. Whether we buy a seat on a spaceship rendezvous to the moon or stay within the confines of an Amish-like commune, we will find no hope apart from our union to Christ. As the center of our lives and our eternal hopes, Christ takes our minds off ourselves and frees us to love and to live for something bigger than our tiny kingdoms. He frees us from slavery to the technological desires of self-creation and self-determining individualism. The church, liberated by Calvin from trying to serve as the arbiter of all scientific discovery, can now preach Christ in the tech age.[134] Only Christ can disenchant the false promises of the tech age. Our gadgets and techno-possibilities no longer define us; Christ does. He defines our calling. If we follow his word, we will be protected from being used by our tools.

In the end, if the farmer, the blacksmith, and Cain's lineage forbid us from dismissing science as a mishap of human arrogance, then we are left with two options: science in faith or faith in science. We choose science in faith. We will listen to the voice of God in creation. *And* we will listen to the voice of God in his Son and in his eternal word. We will listen not only with our ears, but

---

134 Once John Calvin discovered that Christ was the epicenter of Scripture, he freed the church from the role of arbitrating scientific claims, a role Rome attempted to play. Instead, knowing God's central purpose for the Bible as a revelation of Jesus Christ, he focused the church there and freed science to pursue general revelation and to discover the Creator's world. God is at the center of special revelation and general revelation. But the church would specialize on special revelation. See Alister E. McGrath, *A Life of John Calvin: A Study in the Shaping of Western Culture* (Hoboken, NJ: Wiley-Blackwell, 1993), 256.

also with our affections, because the center of the human is not the augmented brain or the superintellect; it's the heart. Because the heart is the center of human existence, we find that there is a "unity between faith, head, hands: unity between faith, science, and technology."[135] Only a cardiocentric worldview can hold together faith, hope, and love with the best offerings of science. Only in Christ can we embrace technological innovation within the healthy confines of present love and future hope.

As Psalm 96 reminds us, God's presence will again physically center on earth. When he does, the entire material universe will sing for joy. Vain idols of power and wealth will be exposed as the foolish worldly shams they are. Instead, the material universe will be brought together to fulfill the ultimate purpose for this creation, the ultimate end of all human innovation: to worship God for the splendor of his beauty and the majesty of his holiness. The sky and earth and seas and fields and forests will rejoice when God himself returns to rule over creation, to rule in perfect justice. All the nations will do the same. And the harmony of the universe will finally flourish in ways we cannot imagine now.

But in this life of the immediate, it is easy to idolize technology and enthrone it as a god. Our hearts are always worshiping, and we will always worship whatever we ultimately think will deliver us. For millions of people, maybe billions of people, that hope is placed in the false god of anti-aging medicine or brain-computer augmentation or genetic self-sculpting or a hundred other corridors of technological innovation. We all know that we must be delivered from our human plight, delivered from ourselves. No one is savior-less. But tech cannot save us.

135 Schuurman, *Faith and Hope in Technology*, 23–24.

As the smoke rises from Babylon, Christ will return to earth in full manifestation of his sovereign power. "From his mouth comes a sharp sword with which to strike down the nations, and he will rule them with a rod of iron. He will tread the winepress of the fury of the wrath of God the Almighty" (Rev. 19:15). He will need no weapons. Nonetheless, razor-sharp war swords, rods of iron, and winepresses—three innovations of man will magnify the power and glory of the returning Christ.

Until Christ returns, we wait. He is our hope. He is the great hope of all true scientists, like the magi who sought in all their designs the moment they would fall down and worship Christ. Until that day, our hope is settled by Paul in Romans 8:18–25. Our sufferings now are small things compared to our hope of that day when Christ restores his creation. The new creation to come is not like this creation, a "creation out of nothing" (*creatio ex nihilo*), but rather a "creation out of the old" (*creatio ex vetere*)—a resurrection, like a dead body made alive. And that is our hope too. Until then, the entire creation "was subjected to futility, not willingly, but because of him who subjected it, in hope that the creation itself will be set free from its bondage to corruption and obtain the freedom of the glory of the children of God" (vv. 20–21).

Creation is the source of our innovations, but creation is not impressed with our gadgets and robotics. Creation groans, like a woman in childbirth. We groan too. We do our best to steward this creation while we await the full redemption of our broken selves, our broken bodies, and this broken creation.

Faith and Physics

I close by returning to where we began, with David's upset over Goliath. It's a story of possibility. A story of faith and physics. A story where trust in God for victory and the right technique for victory are not antithetical. There's no reason that the right technology should nullify our ultimate trust in the delivering power of God. The most authentic wielder of innovation is the one who wields that innovation to love God and love others. Tech is a divine gift to test our stewardship.

As I hope this book has made clear, science cannot deliver us. Innovation will never satisfy our hearts. Life's meaning will never be found in Apple's latest gadget. The God who speaks allows technology to be what it was meant to be—not a savior, not a gospel, and not the final solution to death. Only Christ is the Creator, the meaning of creation, the goal of creation, and the telos of technology. Only Christ is wisdom, the source of wisdom, and the director of innovation used wisely for his ends. Only in him can our hearts exult and our science flourish as the playful craft of humans exploring the generous possibilities of the Creator's material universe.

In Christ, scientists and technologists are freed to cultivate the creation for human flourishing. With a wide-eyed roaming over every possibility, innovators act on that impulse inside us all, that eagerness to discover atoms and unlock power sources and explore planets and travel deeper into space with cameras and then rovers and then ourselves.

Take science too seriously—make it your savior—and it will poison you to the grave. But find your deliverance and joy in the presence of the glorious Savior, and you are in a place to set science free from the serious things of salvation and eternity so that innovation can become the spontaneous and joyous exploration of awe that it was meant to be, the eager study and cultivation of this sandbox we call creation, an intentionally designed gift from the generous Creator we cannot help but worship forever.

Thank-Yous

THIS PROJECT STARTED as a ten-page introduction in my book *12 Ways Your Phone Is Changing You*. Before I could address specific patterns of smartphone use and social media habits, I had to briefly catalogue my meta convictions about human innovation. Over subsequent years, that intro led to many fruitful discussions with friends.

In the summer of 2019, I developed the introduction into a single message that I first shared with Desiring God ministry partners on October 24, 2019—a meta moment, preaching on Babel's tower from the sixth-floor rooftop of the Hotel Valley Ho in Scottsdale, Arizona. Thank you, Sam Macrane!

That message led to an evening and two full days of conversations that led me to write a second and third message. I took the three-part series to three different locations in Seattle in March of 2020. Each gathering brought together insightful pastors and believers working inside the tech industry who offered new feedback. Even though the coronavirus had largely shut down Seattle, we pulled off the events. Thank you, Doug and Mary Lynn Spear and Peter Hedstrom!

Discussions in Seattle made it clear that I needed a theology of the city. So the series grew into four parts, delivered to a group of Christians in Silicon Valley as a live webinar during the lockdown in May. Thank you, Rijo Simon!

As I turned the lectures into a book, I benefited from further engagement with friends, editors, tech experts, and theological brains who invested their time and talents into this project. Thank you, Alastair, Fred, Scott, Lydia, Dilip, Eric, Michael, and Jeremy!

And, finally, this project would be unthinkable without the woman who first became my editor, then my friend, and then my wife. She carries more than her share of life's load to make room for me to write and then also makes more time available in order to edit my prose into something decipherable. To my editor-friend-bride: I love you!

General Index

Abel, 81–83
abortion, 165, 258
Abraham, way of faith, 121
acts of love, 15
Adah, 85
Adam, sin of, 181, 184
Adam and Eve, 80–81, 240–41
adaptation (phase of technological advance), 236
adoption (phase of technological advance), 236
agrarians, 29
agri-business, 96
agricultural technology, 91–94, 99–100
agriculture, as gift of God, 101
AI. *See* artificial intelligence (AI)
Aldrin, Buzz, 154
"already" and "not yet," 121
Amazon Prime, 282
Amish, 24, 43, 269–72, 282
ammonia, 99
analgesics, 176
ancient religions, appeased gods to control nature, 193
anesthetics, 176
angels, 277–78
animal breeding, 86, 87, 89
animals, extinction of, 143
anthrax, 264
anti-aging, 164, 181, 268
antibiotics, 176

antiseptics, 176
Apollo 11 moon landing, 154–55
ark, 89
Armstrong, Neil, 154, 156
Arthur, W. Brian, 31, 97, 126–27
artificial intelligence (AI), 27, 254–60;
    promise of godlikeness, 273–74;
    and providence, 294
aspiration, 41, 227–28
aspirin, 176
atheism, 120, 279
atomic bomb, 59–60, 169
Augustine, 15–16, 264–66
autobots, 58
autonomy: of Babel, 36; as fantasy, 267

Babel, 14, 24, 35–44, 73, 275; attempt to transcend nature, 63; expression of human autonomy, 214, 267; genesis of ethnic animosity, 41–42; got hacked by God, 273; idolatry of, 163; progress thwarted at, 194; as prototype for cities, 120; sarcasm in the story, 36, 44
Babylon, 195–205; capital of self-sufficiency, 197, 199; dehumanized God's image bearers, 201; idolatry of, 200–201, 204–5; opulence of, 204–5, 210, 215; persecution of Christians, 204; power of, 215
Babylonian exile, 207–8
bara, 48, 66, 76, 78, 279

Instagram, 210
intelligent design, of creation, 118
interruption limiter, 293
invasion etiology, 176
iPhones, 69, 97
Iron Age, 86, 88, 90
Israel, flourishing in the promised land, 147

Jabal, 86, 88, 90, 120, 203, 205, 216, 261
Jansen, Kathrin, 110
Job, search for wisdom, 224, 228–31
Jobs, Steve, 68, 69, 75, 88
journalists, 55
Jubal, 86, 88, 90, 120, 203, 205, 216, 261, 277
judgment, on Babel, 39–40

Kay, Alan, 191
Keller, Tim, 206
Kelly, Kevin, 22–23, 24, 26–27, 104, 123, 127, 163, 168, 246, 251, 256, 270, 271, 274–75, 276–77, 289
kinetic energy, 19
Kranzberg, Melvin, 70
Kurzweil, Ray, 170
Kuyper, Abraham, 26, 103, 105–7, 109n63, 125

Lamech, 87
language, 98
Lanier, Jaron, 193–94
Large Hadron Collider, 229
Lee, Kai-Fu, 256, 258
LEGO illustration, 113–15, 118, 124
Lewis, C. S., 169, 170, 185
light bulb, 174–75
light and darkness, 51
lightning, 231
lithium battery farms, 43
loneliness, 282
love, as primary human vocation, 258–60

mariner's compass, 284–85
Mars, experiments on, 251
material world, divorced from spiritual world, 278–79

Matrix (film), 165
matter, does not explain the existence of matter, 78
medical technology, 191; advances because of illness, 106; as soteriology, 165
medicine: healing the sick versus upgrading the healthy, 253; major breakthroughs in nineteenth century, 176
metalworking, 46–48, 89
meteorites, 47
military might, 160–61
mimetic worldview, 233–34
mining techniques (Job), 224–28
miracles, 109–10
mobile housing, 86, 87
Modern Times (film), 280
morphine, 176
multiple discovery, 123
multiplying languages, 40–41
Mumford, Lewis, 227
music industry, 86, 89
Musk, Elon, 27, 43, 65, 67, 68, 156, 165, 235, 252, 256, 280

Naamah, 86
natural laws, 124, 128, 151
natural resources, 128, 131
nature: human desire to transcend, 63; as living organism, 250
navigation, 231
Nazi eugenics, 253
Nebuchadnezzar, 207
Neuralink, 165
new creation, 300
New Jerusalem, 216–22
New Zealand, biodiversity of, 43
Nietzsche, Friedrich, 171
Nineveh, 212
nitrogen supplements, 128
Noah, 14, 34–35, 44, 73, 87, 89, 221
nuclear fission, 232
nuclear fusion, 111, 128, 232

oceans, 231
O'Donovan, Oliver, 241–42
old technologies, merge into new technologies, 126

steam power, 124–25
stewardship of creation, 79, 80, 300
strip mining, 228
subversive weakness, 288
supercomputers, 23
survival of the fittest, in Gospel of Technology, 167
sword makers. *See* blacksmiths
sword wielder. *See* ravager

tar, 33–34, 35, 36, 38, 73
tech-confidence, 161
tech culture, as acceleration culture, 166
tech ethics, not binary, 69–70
tech evolution, 164
tech geniuses, as idolaters, 140
tech limiters, 292–96
tech mastery, requires self-limitation, 270
tech minimalism, 266–72
technical fundamentalism, 172
technicism, 29, 173, 267
technium, 22–24, 276; and discontent, 289; power of, 285–88; and trust in God, 295
techno-agnosticism, 148
techno-autonomy, 146
technocrats, 29
techno-dehumanization, 281, 296
techno-humanism, 27
technological asceticism, 68
technologists, as deacons of discovery, 111
technology: acceleration of, 174–77, 191; advance, 97, 126, 236; advances for the church, 141–43; has ancestors, 97–100; biases of, 210; can never satisfy our souls, 190; cannot solve our greatest need, 151; coded by the Creator into creation, 128; damages creation, 143–45; definition of, 14–17; discontent with, 288–92; displaces gods, 193; displays the glory of God, 246; from dust and returns to dust, 129–30; as gift of God to push back the curse, 132–34; gifts of, 276–79, 301; glorified, 220–22; idolizing of, 146, 189; mastering

of, 43; myths about, 28–29; not morally neutral, 70; in the New Jerusalem, 217–22; now a household term, 17; overreach and corrections of, 247–50; redemption from, 165, 194; remains under the futility of the curse, 134–36; should honor God's design for the body, 250–53; should restore what is broken, 238; stages of, 22
techno-mastery, 120
technopoly, 29
techno-tyranny, 280
tech optimists, 29, 30, 135
tech pessimists, 29, 135
textiles, 86, 87
theism, probabilities reduced by modern science, 186
thermodynamics, 128
thermonuclear war, 248
thinking, 231–32
thorns and thistles, 133
thunder, 231
Torrance, T. F., 107
tower of Babel, 31, 37, 160, 210
tower builder, 53
trade winds, 231
transhumanism, 167–68, 181, 198–99, 213
Trump, Donald, 109, 247
trust: in God, 291, 295–96; in technology, 156, 161–63
Tsiolkovsky, Konstantin, 102
Tubal-cain, 86, 88, 90, 120, 203, 205, 216, 261
Twitter, 125, 210
tyranny of philanthropy, 196

Übermensch, 171, 172, 234
universe: as explosion of God's glory, 77; made from nothing, 229
unnecessary innovation, 261–63
Updike, John, 156–57
"useless class," 257, 260
utopia, 24, 37, 190, 212–13

Vanhoozer, Kevin J., 253
vanity, 275, 296

Scripture Index

Also Available from Tony Reinke

For more information, visit **crossway.org**.